AF823335

Lecture Notes in Control and Information Sciences

Volume 477

Series editors

Frank Allgöwer, Stuttgart, Germany
Manfred Morari, Zürich, Switzerland

Series Advisory Board

P. Fleming, University of Sheffield, UK
P. Kokotovic, University of California, Santa Barbara, CA, USA
A.B. Kurzhanski, Moscow State University, Russia
H. Kwakernaak, University of Twente, Enschede, The Netherlands
A. Rantzer, Lund Institute of Technology, Sweden
J.N. Tsitsiklis, MIT, Cambridge, MA, USA

This series aims to report new developments in the fields of control and information sciences—quickly, informally and at a high level. The type of material considered for publication includes:

1. Preliminary drafts of monographs and advanced textbooks
2. Lectures on a new field, or presenting a new angle on a classical field
3. Research reports
4. Reports of meetings, provided they are
    (a) of exceptional interest and
    (b) devoted to a specific topic. The timeliness of subject material is very important.

More information about this series at http://www.springer.com/series/642

Aram Arutyunov · Dmitry Karamzin
Fernando Lobo Pereira

Optimal Impulsive Control

The Extension Approach

Springer

Aram Arutyunov
Moscow State University
Moscow, Russia

and

Institute of Control Sciences of the Russian Academy of Sciences
Moscow, Russia

and

RUDN University
Moscow, Russia

Dmitry Karamzin
Federal Research Center "Computer Science and Control" of the Russian Academy of Sciences
Moscow, Russia

Fernando Lobo Pereira
FEUP/DEEC
Porto University
Porto, Portugal

ISSN 0170-8643 ISSN 1610-7411 (electronic)
Lecture Notes in Control and Information Sciences
ISBN 978-3-030-02259-4 ISBN 978-3-030-02260-0 (eBook)
https://doi.org/10.1007/978-3-030-02260-0

Library of Congress Control Number: 2018957638

© Springer Nature Switzerland AG 2019, corrected publication 2019
This work is subject to copyright. All rights are reserved by the Publisher, whether the whole or part of the material is concerned, specifically the rights of translation, reprinting, reuse of illustrations, recitation, broadcasting, reproduction on microfilms or in any other physical way, and transmission or information storage and retrieval, electronic adaptation, computer software, or by similar or dissimilar methodology now known or hereafter developed.
The use of general descriptive names, registered names, trademarks, service marks, etc. in this publication does not imply, even in the absence of a specific statement, that such names are exempt from the relevant protective laws and regulations and therefore free for general use.
The publisher, the authors, and the editors are safe to assume that the advice and information in this book are believed to be true and accurate at the date of publication. Neither the publisher nor the authors or the editors give a warranty, express or implied, with respect to the material contained herein or for any errors or omissions that may have been made. The publisher remains neutral with regard to jurisdictional claims in published maps and institutional affiliations.

This Springer imprint is published by the registered company Springer Nature Switzerland AG
The registered company address is: Gewerbestrasse 11, 6330 Cham, Switzerland

Preface

The idea for writing this book emerged from the need to put together a number of research results on optimal impulsive control problems obtained by the authors over the past years. The class of impulsive dynamic optimization problems stems from the fact that many conventional optimal control problems do not have a solution in the classical setting. The absence of a classical solution naturally invokes the so-called extension, or relaxation, of a problem and leads to the concept of generalized solution including the notion of generalized control and trajectory.

Herein, we consider several extensions of optimal control problems within the framework of optimal impulsive control theory. In such a framework, the feasible arcs are permitted to have jumps, and therefore, conventional continuous solutions may fail to exist. Various types of results derived by the authors, essentially centered on the necessary conditions of optimality in the form of Pontryagin's maximum principle, and existence theorems, which shape a substantial body of optimal impulsive control theory, are brought together. At the same time, optimal impulsive control theory is presented in a certain unified framework, while the paradigms of the different problems are introduced in increasing order of complexity. More precisely, the rationale underlying the book consists in addressing extensions of increasing complexity, starting from the simplest case provided by linear control systems and ending with the most general case of totally nonlinear differential control system with state constraints.

This book reflects rather long and laborious cooperative research work undertaken by the authors over the past years, most of which was developed through strong networking between the authors' institutions, the emphasis being on the Research Unit SYSTEC of Engineering Faculty of the University of Porto. Such work certainly would not have been possible, if it were not for the multilateral support of a whole team of researchers and staff. In this regard, we would like to express our gratitude to a number of people who, through their constant research, administrative or logistic support, helped us in our effort to write this book. First of all, we are sincerely grateful to our teachers, co-authors, and colleagues. We especially thank Professor Richard Vinter from Imperial College London. We owe much to the constant assistance of the staff of DEEC/FEUP, notably, Paulo Manuel Lopes, Isidro Ribeiro Pereira, Pedro Lopes Ribeiro, and José António Nogueira. We gratefully acknowledge the proofreading of Alison Goldstraw Fernandes which contributed to the valuable readability of the book. We also thank Oliver Jackson, Manjula Saravanan and Komala Jaishankar from Springer for their excellent support throughout the publishing process.

Finally, we acknowledge the valuable support from various science foundations, both in Russia and Portugal. This work was supported by the Russian Foundation for Basic Research during the projects 16-31-60005 and 18-29-03061, by the Russian Science Foundation during the project 17-11-01168, and by the Foundation for Science and Technology (Portugal) during the projects FCT R&D Unit SYSTEC—POCI-01-0145-FEDER-006933 funded by ERDF | COMPETE2020 | FCT/MEC | PT2020 extension to 2018, NORTE-01-0145-FEDER-000033 funded by ERDF | NORTE 2020, and POCI-01-0145-FEDER-032485 funded by ERDF | COMPETE2020.

Porto, Portugal
July 15, 2018

Aram Arutyunov
Dmitry Karamzin
Fernando Lobo Pereira

Contents

Notation

$\mathbb{R}^k$	Euclidean space of dimension k
$\mathbb{R}$	Real line: $\mathbb{R} = \mathbb{R}^1$
$C(T;\mathbb{R}^k)$	Space of continuous vector-valued functions $x(t) = (x^1(t), x^2(t), \ldots x^k(t))$ defined on T with values in $\mathbb{R}^k$, w.r.t. the norm $\|x\|_C = \max_{t\in T}\lvert x(t)\rvert$
$C^*(T;\mathbb{R}^k)$	Dual space to $C(T;\mathbb{R}^k)$, or the space of Borel vector-valued measures $\mu = (\mu^1, \mu^2, \ldots, \mu^k)$ with values in $\mathbb{R}^k$, w.r.t. the norm of total variation $\|\mu\| = \sup_{\|f\|_C=1} \int_T \langle f, d\mu\rangle$, where notation $\int_T \langle f, d\mu\rangle$ stands for the sum of integrals $\sum_i \int_T f^i d\mu^i$, being f^i the components of f, $i = 1, 2, \ldots, k$
$BV(T;\mathbb{R}^k)$	Space of vector-valued functions $x(t)$ of bounded variation defined on T with values in $\mathbb{R}^k$, w.r.t. the norm $\|x\|_{BV} = \operatorname{Var} x\vert_T$
$L_p(T;\mathbb{R}^k)$	Space of measurable vector-valued functions $x(t)$ defined on T with values in $\mathbb{R}^k$ s.t., $\lvert x(t)\rvert^p$ is integrable, $1 \le p < \infty$, w.r.t. the norm $\|x\|_{L_p} = (\int_T \lvert x(t)\rvert^p dt)^{1/p}$, while for $p = \infty$, $x(t)$ is essentially bounded, w.r.t. the norm $\|x\|_{L_\infty} = \text{ess}\,\sup_{t\in T} \lvert x(t)\rvert$
$W_{1,p}(T;\mathbb{R}^k)$	Space of absolutely continuous vector-valued functions $x(t)$ defined on T with values in $\mathbb{R}^k$, and such that $\frac{dx}{dt} \in L_p(T;\mathbb{R}^k)$
μ	Vector-valued Borel measure
μ_c	Continuous component of measure μ
μ_d	Discrete, or atomic, component of measure μ
μ_{ac}	Absolute continuous component of measure μ
μ_{sc}	Singular continuous component of measure μ
$\operatorname{range}\mu$	Set of all possible values of measure μ
$\operatorname{supp}\mu$	Support of measure μ
$\lvert\mu\rvert$	Total variation measure of measure μ
$\|\mu\|$	Norm given by total variation of measure μ on T, that is $\|\mu\| = \lvert\mu\rvert(T)$
$\mu(\tau)$	Value of measure μ on the set $\{\tau\}$, that is $\mu(\{\tau\})$

$\mathrm{D}s(\mu)$	Set of atoms of measure $\lvert\mu\rvert$
$\mathrm{C}s(\mu)$	Set of points where measure $\lvert\mu\rvert$ is continuous including the endpoints
$F(t;\mu)$	Characteristic function of measure μ
$\mathrm{Var}\, f\rvert_{[a,b]}$	Variation of function f on $[a, b]$
$N_S(p)$	Limiting normal cone to set S at point p
$\partial f(x)$	Limiting subdifferential of function f at point x
$\mathrm{epi}\, f$	Epigraph of function f
$\Pi_S(y)$	Euclidean projection of point y onto the set S
$\mathrm{cone}A$	Conic hull of set A
$\mathrm{conv}A$	Convex hull of set A
$\mathrm{lin}\, A$	Linear hull of set A
$\mathrm{cl}\, A$	Closure of set A
$\mathrm{int}\, A$	Interior of set A
∂A	Boundary of set A
ℓ	Lebesgue measure on $\mathbb{R}$
K°	Polar cone to cone K
$\mathscr{B}(T)$	σ-algebra of Borel subsets of T
$\mathscr{L}(T)$	σ-algebra of Lebesgue subsets of T
M^*	Transposed matrix to M
$\xrightarrow{w}$	Weak convergence of measurable functions (in L_2)
$\xrightarrow{w*}$	Weak-* convergence of measures (in C^* or $\mathrm{B}V$)
$\rightrightarrows$	Uniform convergence of continuous functions (in C)
Limsup	Sequential Painlevé–Kuratowski upper/outer limit
$\mathrm{ind}\Omega$	Index of quadratic form Ω
$f(\tau^-)$	Limit on the left of function f at point τ
$f(\tau^+)$	Limit on the right of function f at point τ
χ_D	Characteristic/indicator function of set D
$\sigma_D(y)$	Support function of set D at point y
B_X	Unit closed ball in space X
$\mathbb{R}^k_+$	Closed positive orthant in $\mathbb{R}^k$
$\mathrm{dist}(y, A)$	Distance from point y to set A
δ_v	Dirac measure concentrated at point $v \in \mathbb{R}^m$
$f \circ g$	Composition of two maps

List of Assertions

Introduction

Abstract The purpose of this chapter is to introduce our book. A brief historical overview is given, the background of the theory of optimal impulsive control with a number of important pieces of work on this theory being pointed out. The key notion of "Extension approach" is disclosed. Finally, for the convenience of reading, the structure of the book and the relationship of its chapters are described. The logic underlying the book consists in moving from simple to complex—starting with the basic linear problem in Chap. 1 and ending with the general nonlinear dynamics under state constraints in Chap. 7.

Some problems of calculus of variations do not have a solution in the classical setting. The following example is among the simplest ones illustrating this property.

$$\begin{aligned} &\text{Minimize} \quad \int_0^1 x^2(t)dt \\ &\text{subject to} \quad x(0) = 0, \ x(1) = 1. \end{aligned} \tag{1}$$

Here, the minimum value of the integral is sought on the set of all smooth functions $x(\cdot)$, whose values at the time endpoints are fixed. A classic smooth solution to problem (1) does not exist. Indeed, there is not even a continuous solution: Any minimizing sequence of arcs for (1) converges pointwise to the discontinuous function $y(\cdot)$ specified by $y(t) = 0$ if $t \in [0, 1)$, and $y(1) = 1$.

Should we consider optimal control problems, this phenomenon takes on a more complicated form for the following reason. There are examples of constrained differential control systems, for which not even a single continuous admissible arc exists. Indeed, on time interval $[0, 1]$, consider the control system

$$\begin{gathered} \dot{x} = u, \ \dot{y} = 1 - x, \ u \geq 0 \\ x(0) = 0, \ x(1) = 1, \ y(0) = y(1) = 0. \end{gathered} \tag{2}$$

It is clear that $x(\cdot)$ is increasing, and $x(t) \leq 1 \ \forall t \in [0,1]$. Therefore, $y(\cdot)$ is increasing as well. Then, in view of the terminal constraints, the compatibility of this system is possible only if $y(t) \equiv 0$. Hence, $x(t) = 1 \ \forall t \in (0,1]$, and, thus, the function $x(\cdot)$ is discontinuous at point $t = 0$. By considering any cost functional over the arcs satisfying (2), we encounter a problem in which continuous solutions fail to exist.

The absence of a classical solution naturally gives rise to the so-called extension, or relaxation, of the problem of calculus of variations or optimal control and leads to the notion of generalized solution. As a matter of fact, the extension of a classic problem stands for the introduction and specification of a certain generalized solution concept: It is necessary to relax the notion of the arc, that is, to enlarge the class of admissible functions $x(\cdot)$, so that, in the enlarged class of arcs, solutions to (1) and (2) would already exist.

A correct analogy is irrational numbers. Since the solution to, for example, the equation $x^2 = 2$ does not exist in the set of rational numbers, the number $\sqrt{2}$ was proposed for use in order to overcome this difficulty. This number is nothing but the generalized solution to this equation. It is also the limit of any approximating sequence of rational solutions to perturbed equation $x^2 = 2 + r$, where $r \in \mathbb{Q}$ and $r \to 0$. Then, the notion of number was extended, while the line of real numbers was completed.

The same logic is traced in calculus of variations, and later in optimal control, where, in some works, the classic smooth arcs were replaced by the arcs of bounded variation. Arcs of bounded variation can exhibit discontinuities, and, consequently, they satisfy the requirements imposed for the generalized solutions to (1) and (2). As was the case with irrational numbers, which at the dawn of their existence caused a lot of mistrust and philosophical disputes among scientists, the discontinuous arcs were also regarded as some kind of anomaly. This anomaly was, of course, known in the classic mechanics and calculus of variations. In the middle of the last century, the development of technology for space exploration raised the need to solve rocket space navigation and control problems in which the jumps of state variables naturally arose. Then, alongside the development of control theory on the whole, the discontinuous arcs solutions to calculus of variations problems began to attract increasing attention from researchers. Gradually, they started to be considered and investigated within the framework of a more general theory, known nowadays as the theory of optimal impulsive control.

Various extensions of the classical calculus of variations have been developed. These extensions have been arranged in diverse functional spaces, not necessarily encompassing discontinuous functions (e.g., in $W_{1,p}$), but historically they can all be traced to the general line of development initiated in the work [25] by Hilbert (see the 20th problem therein). In this famous work, Hilbert expresses confidence that any problem of calculus of variations has a solution, if only the term "solution" is interpreted appropriately. Extensions of calculus of variations or control problems are documented in a huge body of the mathematical literature. Concerning the history of this question, see, for example, sources [18, 26, 36, 53, 57].

In this book, we address the issue of extension related to discontinuous arcs. In modern control theory, the investigation of extensions for constrained optimal control problems is of greater interest than for the problems of calculus of variations, whose formulation is less general. Note that the concept of generalized solution in the optimal control context becomes broader, because together with the notion of generalized arc, the notion of generalized control also arises. Since optimal control, in substance, contains in itself the theory of calculus of variations, the phenomenon of discontinuous arcs is naturally inherited. Moreover, the situation becomes even more complicated for investigation as is illustrated by Example (2). Note that the above-specified property of (2) exhibits rather strict requirements for the construction of approximations by conventional problems with Lipschitz continuous arcs.

Therefore, for the optimal control problem

$$\begin{aligned} &\text{Minimize} && \varphi(x_0, x_1) \\ &\text{Subject to} && \dot{x} = f(x, u, t) \\ &&& x_0 = x(0) \in A, \;\; x_1 = x(1) \in B, \\ &&& u(t) \in U \text{ a.a. } t \in [0, 1], \end{aligned} \tag{3}$$

where φ and f are given smooth functions, A, B, and U are given closed sets, and u is a measurable function (e.g., in the space L_p, $p \geq 1$), the solutions can, generally speaking, be expected to have discontinuities.

At this point, an important clarification has to be made. Discontinuous trajectories can only be expected when the set U is unbounded, or, to be more precise, when the velocity set $f(x, U, t)$ is unbounded for some x, t. Indeed, on the one hand, it is clear that discontinuities occur when the arc derivative takes on unbounded values as in Example (1). Note that this can only happen when the set U is unbounded. On the other hand, in the case of bounded set U, a certain well-established extension of Problem (3) in the class of absolutely continuous arcs already exists under fairly general assumptions. This extension was proposed in [18]. It is based on the notion of a generalized control. Recall that the generalized control is a weakly measurable family of probability Radon measures $\nu_t : \mathscr{B}(U) \to [0, 1]$, $t \in [0, 1]$, where $\mathscr{B}(U)$ stands for the σ-algebra of Borel subsets of U. (For a complete definition, see Chap. 7, where this concept plays a key role.)

If the set U is compact, then the set of generalized controls is weakly sequentially compact [18]. This property of generalized controls is critical for the existence of solutions in the extended problem. However, if the set U is unbounded, then this is not the case. For this reason, when we extend a control problem, it is convenient to split the control parameter into two parts, u and v, and assume that $u \in U$, where set U is compact, and $v \in V$, where set V is closed but not compact (i.e., unbounded, such as a closed cone).

Thus, the idea is to consider solutions to control problems with unbounded velocity set in a generalized sense, or rather, to extend the concepts of control and of trajectory themselves. This procedure is called “extension.” An important characteristic of such a procedure is the property of well-posedness. Well-posedness is an essential requirement for any practical application, with the emphasis on Engineering and Science, and, therefore, it follows that it is important for an extension to be well-posed. A well-posed extension procedure should include the following steps:

1. Extension of controls. Definition of control in the extended problem as an element of a certain metric space. The concept of "proximity" for controls.
2. Extension of trajectories. Definition of trajectory in the extended problem as a solution to the extended dynamical system. The well-posedness of solution consists in verifying that "close" controls yield "close" trajectories.
3. Existence theorems. These are the most important conditions under which the proposed extension procedure is appropriate, in the sense that the extended problem already has a solution in contrast to the original formulation.

In principle, the extension procedure should also preserve the rule: "the original and the extended problems have the same infimum costs." However, this is not always achievable for optimal control problems with constraints; see, for example, Problem (2). This requirement can be achieved through a relaxation of constraints.

Let us look at some types of extensions, starting with the simplest case. On the given time interval $[0,1]$, the measurable control function $v(\cdot)$ can be replaced, for example, by a Borel measure. Indeed, any absolutely continuous Borel measure can be associated with the corresponding integrable function by means of the Radon–Nikodym derivative: $\frac{d\mu}{dt}(t) = v(t)$. However, there are measures that cannot be associated with any measurable integrable function, such as the Dirac measure. It is easy to verify that this approach, using Borel measures, offers an extension for the following dynamic system affine w.r.t. the control variable v:

$$\dot{x} = f(x,u,t) + G(t)v, \quad v \in K, \tag{4}$$

where u is the conventional measurable and essentially bounded control, which takes values in the set U, while v is an unbounded control function, which takes values in a closed convex cone K, and G is a smooth matrix function.

By replacing the function v with the Borel measure μ, and by minimizing the terminal functional of Problem (3) over the solutions to the extended control differential system with measure

$$dx = f(x,u,t)dt + G(t)d\mu,$$

we arrive at the simplest optimal impulsive control problem formulation, in which measure μ designates the *impulsive control*. The extended trajectories x's are, now, functions of bounded variation, and, therefore, they may exhibit jumps.

The idea of such an extension was first presented in [46, 55]. This extension, relying on the weak-* sequential compactness of the unit ball in the space of Borel measures, is well-posed and, in particular, offers a simple existence theorem for solutions (under some additional, but quite natural, assumptions).

The case of affine control systems and the associated well-posed extension is considered in Chap. 1.

In Chap. 2, we consider a more general impulsive control system of the form

$$dx = f(x, u, t)dt + G(u, t)d\mu,$$

involving Borel measurable control $u(\cdot)$ and also the data Borel measurable w.r.t. u. We refer to this case as impulsive control problems under the condition of Borel measurability. A specific approach allows us to overcome the difficulties arising from the fact that the data is merely Borel measurable w.r.t. u. Concerning the Borel measurability of controls, we note that, with regard to the needs of optimal control theory, both types of controls, that is Borel measurable and Lebesgue measurable controls, are equivalent. Indeed, the more general case of Lebesgue measurable controls is easily reduced to the case of Borel measurable controls. (See the discussion in the introduction of Chap. 2.) In the remaining chapters, the data are continuous w.r.t. u and the control functions are Lebesgue measurable. Regarding the solution concept adopted in this chapter, it is the same as in Chap. 1, bringing nothing new as yet to the issue of extension itself.

This, however, is not the case in Chap. 3, where we start to address more sophisticated control dynamic systems

$$\dot{x} = f(x, u, t) + G(x, t)v, \quad v \in K. \tag{5}$$

The difference between (5) and (4) consists only in the fact that the matrix G now depends on the state variable x. The solution concept and the extension procedure in this nonlinear case appears to be more intricate than the one in which G depends only on t. The key point is to ensure robustness or stability of the solution to the impulsive control system w.r.t. the control measure and regarding the approximations/perturbations in the weak-* topology. Note that such approximations are always required by practical applications. This robustness property is achieved by introducing a new concept of solution to the differential equation with measure. This concept extends the one for the simpler case given by (4) when $G = G(t)$. However, it does not rely on the notion of the Lebesgue integral because a different type of integration is used.

Nevertheless, the new definition of solution still fails to satisfy the robustness property unless some extra assumptions on the matrix G w.r.t. the x-variable are imposed. It appears that the most general assumption under which the robustness property is satisfied is the so-called *Frobenius condition*. This condition states that the vector fields G^j—the columns of the matrix G—pairwise commute, that is $(G^j)'_x G^i \equiv (G^i)'_x G^j$ for all i, and j. In Chap. 3, we derive, under the Frobenius condition and without a priori regularity assumptions, the maximum principle and the second-order necessary optimality conditions that remain meaningful even in the abnormal case.

The complexity of the extension significantly increases, should we consider the differential control system (5) without the Frobenius condition. Then, the extension procedure, briefly outlined above for (4), is no longer valid for (5).[1] This is a consequence of the fact that weak-* limits in the nonlinear systems like (5) are not well-posed (i.e., the limit of the corresponding integrals is not equal to the integral of the limit). The incorrectness of such limits can be illustrated with simple examples. Consider the two-dimensional dynamical system with unbounded control $v = (v_1, v_2)$:

$$\dot{x} = xv_1 + x^2 v_2, \quad x(0) = 1. \tag{6}$$

If we try to extend (6) in the class of Borel measures, assuming thereby, that $\|v\|_{L_1} \leq \text{const}$, we can see that to each control, that is, to each vector-valued measure, there is a whole integral funnel of trajectories corresponding to (6). Each one of such trajectories results from a specific approximating sequence of absolutely continuous measures and, therefore, can be called a solution to the dynamic system w.r.t. a given vector-valued measure.

The extension to (5) in the absence of the Frobenius condition and the optimal impulsive control problems related to it are studied in Chaps. 4 and 5. This type of problem has been the subject of interest to many researchers, there being a vast array of publications on this topic. The following sample should not be regarded as exhaustive: [9, 15, 16, 19, 33, 34, 58], and [3, 4, 27].

It follows that, in the case of general system (5), i.e., when the Frobenius condition might not be valid, the space of Borel measures is no longer large enough to encompass all feasible controls of the extended problem, that is, impulsive controls. The impulsive control becomes something more than just a Borel measure and now is already a pair $(\mu; \{v_\tau\})$, where μ is a Borel measure, and $\{v_\tau\}$ is a family of conventional measurable and essentially bounded functions, termed *attached family of controls*. The idea of introducing this attached family consists in dealing with the evolution of the system jumps, expressly, at the time when an impulse occurs. In articles [3, 27], it has been shown that the integral funnel of trajectories arising as a result of the approximation of measure μ by absolutely continuous measures is fully determined by the variety of the attached families to μ. Any attached family of controls defines a certain way of approximation of μ by absolutely continuous measures, and conversely, for any such approximation, there exists a corresponding collection of attached families of controls generating some trajectories of the funnel (see Lemmas 4.1 and 4.2). So, the integral funnel can be parameterized by the attached families. By means of the attached family, we select a path from the integral funnel. This path becomes the solution, corresponding to the given pair—measure plus attached family. In other words, the attached family of controls is nothing but a method of approximation by absolutely continuous measures. Figuratively speaking, it is “an interaction scheme of the components of the vector-valued measure at the moment of impulse.”

[1] More precisely, it is applicable, but it does not yield a well-posed extension. Therefore, it does not offer an extension interesting from the point of view of applications.

Note that under the Frobenius condition the introduction of attached family of controls is redundant, because the above-mentioned integral funnel degenerates into one single arc (see, e.g., [10, 15, 58], and also Excercise 4.2). Therefore, while the extension considered in Chap. 3 can still be described using Borel measures as impulsive controls, the extensions of the ensuing chapters require a significant relaxation of the space of Borel measures $C^*([0, 1]; \mathbb{R}^k)$.

It is not surprising that an even more difficult case is the one for which the matrix G depends not only on the state variable x, but also on the control variable u, that is, when the dynamic control system is specified by

$$\dot{x} = f(x, u, t) + G(x, u, t)v, \quad v \in K. \tag{7}$$

In this case, the introduction of additional controls acting on the jumps of trajectories is necessary, even if the Frobenius condition is satisfied. Extensions for (7) have been studied, for example, in works [7, 8, 20, 29, 33, 34]. Note that, despite the increasing complexity of the extension procedure, the dependence of G on the bounded control variable u is not a mere mathematical generalization. Control systems such as (7) can be encountered in various engineering applications, in which it is important, for example, to take into consideration the rapid variations in the mass distribution of a mechanical system over the small time interval when the impulse occurs (see, e.g., [8, 29]).

Chapter 6 is devoted to the study of case (7) where we also discuss a model example of a mechanical system of the above-specified type.

Finally, we consider the most general case

$$\dot{x} = g(x, v, t), \quad v \in V. \tag{8}$$

Here, g is some smooth vector-valued function, and the set V is, as it was noted earlier, closed and unbounded. System (8), obviously, includes the cases of (3)–(5), and (7). Various extensions for this general case given by (8) have been studied, e.g., in [28, 35, 38, 47, 55].

Several approaches exist in the literature on how to deal with the case of control dynamics (8). The idea proposed in [28] consists in bonding the Borel measure μ defined over $[0, 1]$ and the generalized controls in the sense of Gamkrelidze defined over V, by using the Lebesgue discontinuous time variable change. Such a composition of different types of controls, coupled with a standard compactification procedure, leads to a rather general extension of the original problem in the class of problems with trajectories of bounded variation.

The extension for (8) is the subject of Chap. 7.

Overall, there is a huge array of publications on the subject of impulsive control theory. Besides the work referred to above, the contributions made in [1, 2, 5, 11, 14, 21–24, 30–32, 37, 39, 40, 42–44, 48, 50–52, 56] could also be pointed out, while this selected list can be complemented with many other pieces of work. However, in this book, the aim is not to provide a full-fledged review encompassing the whole range of approaches to impulsive control.

It is clear from what has been said above that the logic behind this book consists in addressing the increasing complexity for extensions, starting from the simplest case with the control dynamics given by (4) and ending with the most general case, provided by (8). Throughout all seven chapters, (4), (5), (7), and (8) are treated sequentially by increasing complexity of the models.

However, the problem of extension is not the only area of interest in this book. An important part of optimal control theory is constituted by various types of constraints imposed on the control process, or rather on the pair trajectory plus control. Several basic types of constraints have been considered in the literature: terminal (or endpoint) constraints and state and mixed constraints. With regard to constraints, the outline of the book is as follows. In all chapters, the geometric endpoint constraints given by the inclusion $(x(0), x(1)) \in S$, where S is a closed set, are considered. Only in those sections of Chap. 3 devoted to second-order optimality conditions, these constraints are specified functionally. In Chap. 5, we additionally consider inequality state constraints of the form $l(x,t) \leq 0$. In Chap. 6, a mixed constrained impulsive control problem is investigated; that is, the optimal control problem formulation includes inequality mixed constraints of the form $r(x,u,t) \leq 0$. In Chap. 7, we return to the state constrained problems. This is justified by a series of classic examples to which the theory of Chap. 7 is applied. These essentially involve state constraints.

Among all types of constraints, the most difficult in the study are, obviously, state constraints. Despite the fact that, formally, state constraints can be regarded as a mere particular case of the mixed constraints, the consideration of state constraints encounters many more difficulties for the following simple reason. The mixed constraints are commonly assumed to be regular w.r.t. the control variable. This imposes a certain normality assumption on the derivative of r w.r.t. u, and, in particular, it prohibits this derivative from vanishing. However, the derivative of l w.r.t. u vanishes everywhere because l does not depend on u at all. Therefore, the state constraints belong to the category of nonregular mixed constraints. This fact of involved nonregularity expressly explains the difficulties arising in the study of optimal control problems with state constraints.

In this book, the Gamkrelidze approach, [17], for the investigation of the necessary optimality conditions in problems with state constraints is used (see also [6, 41, 45, 49, 54]). This approach requires the existence of the second-order derivative of l, the function specifying the state constraints. This extra requirement serves for various useful properties of the Lagrange multipliers and the maximum principle, and for applications. This topic is discussed in detail in Chap. 5. Here, it is sufficient to note that, if the second-order derivative of l exists, the Gamkrelidze form of the maximum principle is equivalent to the Dubovitskii–Milyutin form, [12, 13], due to a variable change in the multipliers (see in [6, 41]).

Each chapter is enriched with a number of exercises. These exercises are intended to help with the perception of the new concepts introduced in this book, as well as to shed more light on some of the technical results. Some of the exercises are rather simple and constitute tasks for graduate or postgraduate students. The others are of medium and high difficulty. High-difficulty tasks are marked with the asterisk.

References

1. Aronna, M.S., Motta, M., Rampazzo, F.: Infimum gaps for limit solutions. Set-Valued Var. Anal. **23**(1), 3–22 (2015)
2. Arutyunov, A., Jacimovic, V., Pereira, F.: Second order necessary conditions for optimal impulsive control problems. J. Dyn. Control Syst. **9**(1), 131–153 (2003)
3. Arutyunov, A., Karamzin, D.: Necessary conditions for a minimum in optimal impulse control problems. In: Yemelyanov, S.V., Korovin, S.K. (eds.) Nonlinear Dynamics and Control Collection of Articles. Fizmatlit, Moscow **4**(5), 205–240 (2004) (in Russian)
4. Arutyunov, A., Karamzin, D., Pereira, F.: A nondegenerate maximum principle for the impulse control problem with state constraints. SIAM J. Control Optim. **43**(5), 1812–1843 (2005)
5. Arutyunov, A., Karamzin, D., Pereira, F.: Pontryagin's maximum principle for optimal impulsive control problems. Doklady Math. **81**(3), 418–421 (2010)
6. Arutyunov, A., Karamzin, D., Pereira, F.: The maximum principle for optimal control problems with state constraints by R.V. Gamkrelidze: revisited. J. Optim. Theory Appl. **149** (3), 474–493 (2011)
7. Arutyunov, A., Karamzin, D., Pereira, F.: On a generalization of the impulsive control concept: controlling system jumps. Discret. Contin. Dyn. Syst. **29**(2), 403–415 (2011)
8. Arutyunov, A., Karamzin, D., Pereira, F.: Pontryagin's maximum principle for constrained impulsive control problems. Nonlinear Anal. Theory Methods Appl. **75**(3), 1045–1057 (2012)
9. Bressan, A., Rampazzo, F.: On differential systems with vector-valued impulsive controls. Boll. Un. Matematica Italiana 2-B pp. 641–656 (1988)
10. Bressan, A., Rampazzo, F.: Impulsive control systems with commutative vector fields. J. Optim. Theory Appl. **71**(1), 67–83 (1991)
11. Code, W., Loewen, P.: Optimal control of non-convex measure-driven differential inclusions. Set-Valued Var. Anal. **19**(2), 203–235 (2011)
12. Dubovitskii, A., Milyutin, A.: Extremum problems with constraints. Sov. Math. Dokl. **4**, 452–455 (1963)
13. Dubovitskii, A., Milyutin, A.: Extremum problems in the presence of restrictions. USSR Comput. Math. Math. Phys. **5**(3), 1–80 (1965)
14. Dykhta, V.: The variational maximum principle and second-order optimality conditions for impulse processes and singular processes. Siberian Math. J. **35**(1), 65–76 (1994)
15. Dykhta, V., Samsonyuk, O.: Optimal Impulse Control with Applications. Fizmatlit, Moscow (2000) (in Russian)
16. Dykhta, V., Samsonyuk, O.: A maximum principle for smooth optimal impulsive control problems with multipoint state constraints. Comp. Math. Math. Phys. **49**(6), 942–957 (2009)
17. Gamkrelidze, R.: Optimal control processes for bounded phase coordinates. Izv. Akad. Nauk SSSR. Ser. Mat. **24**, 315–356 (1960)
18. Gamkrelidze, R.: Principles of Optimal Control Theory. Plenum Press, New York (1978)
19. Goncharova, E., Staritsyn, M.: Optimization of measure-driven hybrid systems. J. Optim. Theory Appl. **153**(1), 139–156 (2012)

20. Goncharova, E., Staritsyn, M.: Optimal impulsive control problem with state and mixed constraints: The case of vector-valued measure. Autom. Remote Control **76**(3), 377–384 (2015)
21. Goncharova, E., Staritsyn, M.: Approximation and relaxation of mechanical systems with discontinuous velocities. Cybern. Phys. **6**(4), 215–221 (2017)
22. Guerra, M., Sarychev, A.: Existence and lipschitzian regularity for relaxed minimizers (2008)
23. Guerra, M., Sarychev, A.: Frechet generalized trajectories and minimizers for variational problems of low coercivity. J. Dyn. Control Syst. **21**(3), 351–377 (2015)
24. Gusev, M.: On optimal control of generalized processes under nonconvex state constraints. Differential Games and Control Problems. UNTs, Akad. Nauk SSSR, Sverdlovsk **15**, 64–112 (1975) (in Russian)
25. Hilbert, D.: Mathematical problems. Bull. Am. Math. Soc. **8**, 437–479 (1902)
26. Ioffe, A., Tikhomirov, V.: Extension of variational problems. Moscow Math. Soc. Surv. **18**, 187–246 (1968)
27. Karamzin, D.: Necessary conditions of the minimum in an impulse optimal control problem. J. Math. Sci. **139**(6), 7087–7150 (2006)
28. Karamzin, D., De Oliveira, V., Pereira, F., Silva, G.: On some extension of optimal control theory. Eur. J. Control **20**(6), 284–291 (2014)
29. Karamzin, D., De Oliveira, V., Pereira, F., Silva, G.: On the properness of an impulsive control extension of dynamic optimization problems. ESAIM—Control, Optim. Calc. Var. **21**(3), 857–875 (2015)
30. Krotov, V.: Discontinuous solutions of variational problem. Izv. Vyssh. Uchebn. Zaved. Mat. (5), 86–98 (1960) (in Russian)
31. Kurzhanski, A., Daryin, A.: Dynamic programming for impulse controls. Ann. Rev. Control **32**(2), 213–227 (2008)
32. Marec, J.: Optimal Space Trajectories. Studies in Astronautics. Elsevier Scientific Publishing. Co., Amsterdam, Netherlands (1979)
33. Miller, B.: Generalized solutions in nonlinear optimization problems with impulse controls. i. the solution existence problem. Avtomatika i Telemekhanika (4), 62–76 (1995)
34. Miller, B.: Generalized solutions in nonlinear optimization problems with impulse controls. ii. representation of solutions by differential equations with measure. Avtomatika i Telemekhanika (5), 56–70 (1995)
35. Miller, B.: The generalized solutions of nonlinear optimization problems with impulse control. SIAM J. Control Optim. **34**(4), 1420–1440 (1996)
36. Mordukhovich, B.: Existence of optimum controls. J. Sov. Math. **7**(5), 850–886 (1977)
37. Motta, M., Rampazzo, F.: Dynamic programming for nonlinear systems driven by ordinary and impulsive controls. SIAM J. Control Optim. **34**(1), 199–225 (1996)
38. Motta, M., Rampazzo, F.: Nonlinear systems with unbounded controls and state constraints: a problem of proper extension. Nonlinear Differ. Equ. Appl. **3**(2), 191–216 (1996)
39. Murray, J.: Existence theorems for optimal control and calculus of variations problems where the states can jump. SIAM J. Control Optim. **24**(3), 412–438 (1986)
40. Neustadt, L.: A general theory of minimum-fuel space trajectories. J. Soc. Ind. Appl. Math. Ser. A Control **3**(2), 317–356 (1965)
41. Neustadt, L.: An abstract variational theory with applications to a broad class of optimization problems. ii. applications. SIAM J. Control **5**(1), 90–137 (1967)
42. Pereira, F.: A maximum principle for impulsive control problems with state constraints. Comput. Math. Appl. **19**(2), 137–155 (2000)
43. Pereira, F., Silva, G.: Necessary conditions of optimality for vector-valued impulsive control problems. Syst. Control Letters **40**(3), 205–215 (2000)
44. Pereira, F., Silva, G.: Stability for impulsive control systems. Dyn. Syst. **17**(4), 421–434 (2002)

45. Pontryagin, L., Boltyanskii, V., Gamkrelidze, R., Mishchenko, E.: Mathematical theory of optimal processes, 1st edition. Translated from the Russian ed. by L.W. Neustadt. Interscience Publishers, Wiley 1962)
46. Rishel, R.: An extended pontryagin principle for control systems, whose control laws contains measures. J. SIAM. Ser. A. Control **3**(2), 191–205 (1965)
47. Rockafellar, R.: Dual problems of lagrange for arcs of bounded variation. Calculus of variations and control theory. pp. 155–192 Academic Press, New York (1976)
48. Rockafellar, R.: Optimality Conditions for Convex Control Problems with Nonnegative States and the Possibility of Jumps. Game Theory Math. Economics, pp. 339–349. North-Holland, Amsterdam (1981)
49. Russak, I.: On general problems with bounded state variables. J. Optim. Theory Appl. **6**(6), 424–452 (1970)
50. Silva, G., Vinter, R.: Measure driven differential inclusions. J. Math. Anal. Appl. **202**(3), 727–746 (1996)
51. Sorokin, S., Staritsyn, M.: Feedback necessary optimality conditions for a class of terminally constrained state-linear variational problems inspired by impulsive control. Numer. Algebra Control Optim. **7**(2), 201–210 (2017)
52. Vinter, R., Pereira, F.: Maximum principle for optimal processes with discontinuous trajectories. SIAM J. Control Optim. **26**(1), 205–229 (1988)
53. Warga, J.: Relaxed variational problems. J. Math. Anal. Appl. **4**(1), 111–128 (1962)
54. Warga, J.: Minimizing variational curves restricted to a preassigned set. Trans. Amer. Math. Soc. **112**, 432–455 (1964)
55. Warga, J.: Variational problems with unbounded controls. J. SIAM. Ser. A. Control **3**(2), 424–438 (1965)
56. Wolenski, P., Zabic, S.: A sampling method and approximation results for impulsive systems. SIAM J. Control Optim. **46**(3), 983–998 (2007)
57. Young, L.: Generalized curves and the existence of an attained absolute minimum in the calculus of variations. C. R. Soc. Sci. et Lettres Varsovie, Cl. III **30**, 212–234 (1937)
58. Zavalischin, S., Sesekin, A.: Impulse Processes: Models and Applications. Nauka, Moscow (1991) (in Russian)

Chapter 1
Linear Impulsive Control Problems

Abstract In this chapter, the simplest impulsive extension of a control problem which is feasible in the case of linear dynamical control systems is described. The chapter begins by considering several typical examples of linear control problems for which the appearance of discontinuities in admissible trajectories is natural, since it fits into their physical representation (under certain assumptions made from the point of view of the mathematical model). In particular, the well-known Lawden's problem of the motion of a rocket is examined here and it is demonstrated how discontinuities of extremal trajectories inevitably arise. Next, we give a theorem on the existence of a solution to the extended problem and another theorem concerning necessary optimality conditions in the form of Pontryagins maximum principle, which, in the linear case, are expressed in a sufficiently simple and clear way. The chapter ends with 11 exercises.

1.1 Introduction

The purpose of this chapter is introductory, as the simplest case of linear impulsive control systems is under study. It shows how impulsive controls arise by implicating linear examples. It follows that the consideration of linear impulsive control systems is rather straightforward relative to the issues raised in this book, as one can use simple and demonstrative methods to prove the existence of solutions and the necessary optimality conditions in the form of the *Pontryagin maximum principle*, [8]. The existence of solution is well ensured for linear impulsive control problems under some standard compactness hypotheses. The necessary optimality conditions also take rather comprehensible form, while the main differences w.r.t. the conventional maximum principle are already fairly well outlined. At the same time, linear impulsive control systems are already meaningful for simple applications.

Below, a few simple control problems are examined. These uncover and illustrate the origin of impulsive controls.

© Springer Nature Switzerland AG 2019

A. Arutyunov et al., *Optimal Impulsive Control*, Lecture Notes in Control and Information Sciences 477, https://doi.org/10.1007/978-3-030-02260-0_1

Example 1.1 Consider the problem

$$\begin{aligned}&\text{Minimize} \int_0^1 u(t)dt\\&\text{subject to: } \dot{x} = x + u,\\&\qquad x(0) = 0, \quad x(1) = 1, \quad u(t) \geq 0.\end{aligned} \tag{1.1}$$

Here, the scalar x reflects the state of a certain resource—biological, financial, etc.—whose growth, under normal conditions and without any external intervention, is exponential in time, that is, $\dot{x} = x$. For example, such a model is naturally considered for population of microorganisms (cells, bacteria, etc.). However, there is also some control u which, in this case, stands for the regulation of the extra growth for population x. The extra growth can be caused by various reasons, including artificial factors. The important features are the facts that (a) the extra growth factor can increase the population instantaneously, i.e., within a very short time, and (b) the extra growth is related to consumption of a certain extra resource which would be useful to optimize. Let us designate this extra resource by w. Note that due to (a) the control parameter u is allowed to take unbounded values, while, due to (b), it has the meaning of the consumption rate of the extra resource, and thereby, $\dot{w} = -u$. What is the best strategy for bringing the state of resource x from amount 0 to 1 over the time interval $[0, 1]$ with minimal extra resource costs $w(0)$?

The solution to this problem is, of course, evident from the point of view of common sense, and no theory is needed to give it: The best strategy is to spend all the extra resource at the initial time $t = 0$ bearing in mind its future exponential growth.

Nevertheless, we approach this statement from the pure mathematical point of view. Let us show that a classical continuous solution to (1.1) does not exist. Pick positive integer $n = 1, 2, 3, \ldots$ and impose in Problem (1.1) the additional constraint on control variable: $u \in [0, n]$. Then, it is easy to check that, for each n, there exists a feasible trajectory to this reduced n-problem. Therefore, there exists a solution. Denote it by (x_n, u_n). It is clear that the sequence of controls u_n is minimizing for the integral functional in (1.1). Let us calculate u_n by applying the maximum principle. Since $H(x, u, \psi, \lambda) = \psi(x + u) - \lambda u$, the following conditions arise:

$$\dot{\psi}_n = -\psi_n, \quad \max_{[0,n]} u(\psi_n(t) - \lambda_n) = u_n(t)(\psi_n(t) - \lambda_n) \ \text{ a.a. } t \in [0, 1], \quad \lambda_n \geq 0,$$

where ψ_n, λ_n are the Lagrange multipliers for n-problem. Then, $\psi_n(t) = c_n e^{-t}$, and, as is easy to verify, $c_n > 0$, while the optimal control takes the form:

$$u_n(t) = \begin{cases} n, & \text{when } t \in [0, t_n],\\ 0, & \text{otherwise.}\end{cases}$$

The number $t_n \in (0, 1)$ is computed by solving the boundary value problem on x_n as follows. It holds that

$$x_n(t) = \begin{cases} ne^t - n, & \text{when } t \in [0, t_n], \\ (ne^{t_n} - n)e^{t-t_n}, & \text{when } t > t_n \end{cases}$$

Then, in view of $x_n(1) = 1$, we have

$$t_n = \ln \frac{ne}{ne - 1}, \quad n = 1, 2, \ldots$$

At the same time, $\|u_n\|_{L_1} = nt_n = n \ln \frac{ne}{ne-1} \to e^{-1}$, and $x_n(t_n) = \frac{n}{ne-1} \to e^{-1}$. So, the minimal value of the integral converges to e^{-1}, while the sequence of trajectories x_n converges pointwise to the discontinuous function $x(\cdot)$: $x(0) = 0, x(t) = e^{t-1}$ for $t > 0$. This proves that a classical solution to Problem (1.1) does not exist. Note that $u_n(t) \to 0$ for a.a. t, but $\|u_n\|_{L_1} \nrightarrow 0$. This, along with the step form of u_n, suggests that u_n converges (in some sense) to a Dirac function which follows to be a solution to this problem. This Dirac function is concentrated at point $t = 0$ and has the weight of e^{-1}. Then, the limiting trajectory $x(\cdot)$ has the jump at $t = 0$ on this value of e^{-1} and is optimal.

Consider a more sophisticated example. The equation of motion of a rocket has the following form, [4, 5],

$$\begin{aligned} &\dot{\mathbf{x}} = \mathbf{v}, \\ &\dot{\mathbf{v}} = \mathbf{g}(\mathbf{x}) + \tfrac{\mathbf{c}}{M} \cdot \tfrac{dM}{dt}, \end{aligned}$$

where $\mathbf{x}$ is the rocket position, $\mathbf{v}$ is the rocket velocity, M is its mass, $\mathbf{c}$ is the exhaust velocity, and $\mathbf{g}$ is the force field per unit mass. By simplifying the reality, linearizing the gravity force and considering just one dimension, we may state the following example.

Example 1.2 Consider the problem

$$\begin{aligned} \text{Minimize } & \int_0^1 u(t)dt \\ \text{subject to: } & \dot{x} = v, \\ & \dot{v} = -g + x + u, \\ & x(0) = a, \quad x(1) = b, \quad v(0) = c, \\ & g > 0, \quad a < b, \quad u(t) \geq 0. \end{aligned} \tag{1.2}$$

This is a simplified orbital maneuvering problem. The point is to minimize the fuel consumption in the transfer of a spacecraft from orbit a to orbit b with starting velocity c. Here, x is the spacecraft position, v is its velocity, $\mathbf{g} = -g + x$ is the local gravity force field, and control u has the meaning of thrust or fuel consumption rate (as it is assumed that they are proportional). Control u is considered unbounded because the process of fuel consumption in the jet engine has an explosive character, as a considerable amount of fuel can be spent within seconds.

The classical continuous solutions also do not exist in this example. Indeed, by employing the same arguments as in Example 1.1, we impose an additional constraint on control: $u \in [0, n]$, where $n = 3, 4, \ldots$, and consider, for definiteness, that $a = 0$, $b = 1, c = 0, g = 1$. Denote the solution of n-problem by (u_n, x_n, v_n). The application of the maximum principle yields the same step form of u_n as in Example 1.1, where number t_n is computed as the least positive satisfying the following equation:

$$t_n = \frac{1}{2}\ln\Big(\frac{-\alpha_n e^2}{\beta_n}\Big), \quad \text{where}$$

$$\alpha_n = \frac{(n-1)\cdot\Big(\sinh t_n + \cosh t_n - 1\Big) - 1}{2},$$

$$\beta_n = \frac{(n-1)\cdot\Big(\cosh t_n - \sinh t_n - 1\Big) - 1}{2}.$$

The numerical procedure yields the following table.

n	t_n	$n \cdot t_n$	α_n	β_n	$\alpha_n - \beta_n$
10	0.144539	1.44539	0.19978	−1.10561	1.30539
20	0.0686972	1.37394	0.17556	−1.13071	1.30627
50	0.0267265	1.33633	0.16363	−1.14613	1.30975
100	0.0132451	1.32451	0.16000	−1.15131	1.31131
200	0.00659367	1.31873	0.15824	−1.15391	1.31215
500	0.00263061	1.31531	0.15720	−1.15547	1.31268
1000	0.00131417	1.31417	0.15686	−1.15600	1.31285
5000	0.000262652	1.31326	0.15659	−1.15641	1.31300
10000	0.000131315	1.31315	0.15655	−1.15647	1.31302
100000	0.0000131305	1.31305	0.15652	−1.15651	1.31303
1000000	0.0000013130	1.31304	0.15652	−1.15652	1.31304

As the table clearly shows, $t_n \to 0$, the numbers α_n, β_n also converge to some α, β that can be found numerically, while $\|u_n\|_{L_1} = n \cdot t_n$ converges to some number $f_{\min}$ equal to the amount of fuel consumed over the maneuver which, as was proven above, is the minimum possible w.r.t. any other feasible maneuvering. Then, functions $u_n(\cdot)$ converge (in some sense) to a Dirac function concentrated at point $t = 0$ with the weight $f_{\min}$. Functions $v_n(\cdot)$ converge pointwise to discontinuous function $v(\cdot)$ such that $v(0) = 0$ and $v(t) = \alpha e^t - \beta e^{-t}$ for $t \in (0, 1]$ and which is optimal velocity in Problem (1.2). Also, notice that $f_{\min} = \alpha - \beta = v(0^+)$ as the table outlines. Thus, the optimal maneuver is a single thrust in the initial point with the subsequent inertial motion (free fall). A situation very similar to the one from Example 1.1 arose.

Let us complexify the previous example slightly.

Example 1.3 Consider the problem

$$\begin{aligned}
&\text{Minimize} \int_0^1 u(t)dt \\
&\text{subject to: } \dot{x} = v, \\
&\qquad \dot{v} = -g + x + u, \\
&\qquad x(0) = a, \quad x(1) = b, \quad v(0) = c, \quad v(1) = d, \\
&\qquad g > 0, \quad a < b, \quad u(t) \geq 0.
\end{aligned} \tag{1.3}$$

It is merely the same Example 1.2, but the endpoint velocity is now fixed. How will the optimal velocity change?

The answer to this question can be given in view of similar arguments. For reduced n-problem with the extra constraint $u \in [0, n]$, as consequence of the maximum principle, the optimal control acquires the new step-down-step-up shape as follows:

$$u_n(t) = \begin{cases} n, & \text{when } t \in [0, t_{0,n}] \cup [t_{1,n}, 1], \\ 0, & \text{otherwise,} \end{cases}$$

where $0 < t_{0,n} < t_{1,n} < 1$. The numerical experiment will show that $t_{0,n} \to 0$, and $t_{1,n} \to 1$, and, thereby, the optimal maneuver consists now of the two thrusts: one at the starting and the other one at the terminal points with the inertial motion between them.

Now, in order to give a strict mathematical meaning to Problems (1.1)–(1.3) raised in these examples, it is necessary to describe the class of feasible controls. If one of the spaces L_p, $p \geq 1$, is selected to be a candidate for such a class of controls, then, as it has just been illustrated, the infimum value of the minimizing functional in (1.1)–(1.3) is not reached in the desired class, while any minimizing sequence of feasible trajectories converges to a discontinuous function. (See also Exercise 1.1.)

This phenomenon has the following simple explanation. On the one hand, in L_p, for $p > 1$, the set of feasible controls in Problems (1.1)–(1.3) is unbounded. It is bounded only in L_1-norm which is minimized. On the other hand, the unit ball in L_1 is not a weakly sequentially compact set in contrast with the unit ball in the reflexive space L_p, i.e., when $1 < p < \infty$. Therefore, the existence of a solution in any of the classes L_p, $p \geq 1$, cannot be guaranteed by means of conventional theory. The feasible arcs tend to exhibit discontinuities and thereby to fall into a larger class of arcs of bounded variation. In order to find a solution in such degenerate cases, it is necessary to arrange a procedure of extension or relaxation of the problem. This procedure has been briefly outlined and discussed in Introduction. Then, the extended class of controls is said to be impulsive. The main question is: in which topology (metric) should the convergence of controls be understood, or, in other words, w.r.t.

which topology should the extension be performed? By examining linear control problems in this chapter, such an extension procedure is given in its simplest guise.

1.2 Problem Statement

Consider the following problem:

$$\begin{aligned}
&\text{Minimize} \quad \varphi(p)\\
&\text{subject to:} \quad dx = \Big(A(t)x + B(t)u + f(t)\Big)dt + C(t)d\mu,\\
&\qquad x(t_0) = x_0,\ x(t_1) = x_1,\ t \in T = [t_0, t_1],\\
&\qquad p = (x_0, x_1) \in S,\\
&\qquad u(t) \in U \ a.e.,\\
&\qquad \mu \geq 0.
\end{aligned} \tag{1.4}$$

Here, $t \in \mathbb{R}^1$ stands for the time variable, $T = [t_0, t_1]$ for the time interval which is assumed to be fixed (problems with nonfixed time are treated later in the next chapters), $x = (x^1, \ldots, x^n)$ is the state variable, with values in $\mathbb{R}^n$. The vector $p = (x_0, x_1)$ is called *endpoint vector*. Its range is within a given closed set S which defines the endpoint constraints. The vector $u = (u^1, u^2, \ldots, u^m)$ from $\mathbb{R}^m$ is the so-called *control parameter*. The given set $U \subseteq \mathbb{R}^m$ is convex and compact. The measurable and essentially bounded function $u : T \to U$ is called *control function*, or just control, in Problem (1.4). The given matrix-valued functions $A : T \to \mathbb{R}^{n\times n}$, $B : T \to \mathbb{R}^{m\times n}$ are measurable and essentially bounded, while the given matrix-valued function $C : T \to \mathbb{R}^{k\times n}$ is continuous. The given vector-valued function $f : T \to \mathbb{R}^n$ is measurable and essentially bounded. The scalar function $\varphi : \mathbb{R}^{2n} \to \mathbb{R}$ is smooth and is minimized over the pairs $(x(t_0), x(t_1))$, where $x(\cdot)$ is the solution to the differential system in (1.4) for a given control policy.

The main differences in comparison with the classical theory of optimal control become evident at this point. For the reader unfamiliar with the impulsive control theory, the following two questions would be of most interest:

(a) What is μ?
(b) What does the relation "$dx = \ldots$" stand for?

In Problem (1.4), μ defines the *impulsive control parameter* due to which the trajectories $x(\cdot)$'s are allowed to have discontinuities. To be more specific, μ is a vector-valued Borel measure on $T = [t_0, t_1]$, that is, the element of space $C^*(T; \mathbb{R}^k)$. As is known, $\mu = (\mu^1, \mu^2, \ldots, \mu^k)$, where μ^i, $i = 1, 2, \ldots, k$, is a scalar Borel measure defined on T, that is, a σ-additive function of a set defined over the σ-algebra $\mathscr{B}(T)$ of the Borel subsets B of the time interval T, [3]. Scalar measure $\nu \in C(T; \mathbb{R})$ is said to be nonnegative, $\nu \geq 0$, provided that $\nu(B) \geq 0 \ \forall\, B \in \mathscr{B}(T)$. For the vector-valued measure μ, we denote $\mu \geq 0$, provided that $\mu^i \geq 0 \ \forall\, i$. So, in Problem (1.4), it is required that the control μ is a vector-valued Borel measure with nonnegative components. Note that, if $\mu \geq 0$, then

$$\|\mu\| = \sum_{i=1}^{k} \int_{T} d\mu^i = \sum_{i=1}^{k} \mu^i(T).$$

Now, the relation "$dx = \ldots$" means the following: The trajectory $x(\cdot)$ is called the solution to the differential system in (1.4), corresponding to the control u and to the impulsive control μ, if

$$x(t) = x_0 + \int_{t_0}^{t} \Big(A(s)x(s) + B(s)u(s) + f(s)\Big)ds + \int_{[t_0,t]} C(s)d\mu, \quad \forall t > t_0, \tag{1.5}$$

and $x(t_0) = x_0$. Thus, by selecting a function $u(\cdot)$ and a measure μ, we obtain a single trajectory $x(\cdot)$ which is not now necessarily continuous: It may have jumps whenever measures μ^j have atoms. Function $x(\cdot)$ defined by (1.5) belongs to a space larger than that of continuous functions, expressly, to the space of functions of bounded variation: $x \in \mathrm{BV}(T; \mathbb{R}^n)$.

This phenomenon of phase trajectories exhibiting jumps is not something exceptional but is rather common and often encountered in physics. Indeed, the velocity in mechanical problems is considered as a state variable component. However, the velocity is discontinuous in, e.g., the impact mechanics. The simplest example is given by the collision of two balls when the velocity changes its direction and absolute value at the moment of impact. Mathematically, the moment of collision may be modeled by means of Borel measures which can have atoms and, can, therefore, instantaneously change the value of the velocity.

In Problem (1.4), the cost is given in the form of the so-called terminal or endpoint functional $\varphi(p)$. However, by introducing a new state variable y, such that $dy = \langle c, d\mu \rangle$, $y(t_0) = 0$, we can always proceed to the integral functional

$$y(t_1) = \int_{[t_0,t_1]} \langle c, d\mu \rangle \to \min . \tag{1.6}$$

This integral form (1.6) of the cost is crucial for many actual applications. Indeed, as discussed in Examples 1.1–1.3, the situation in which the trajectory becomes unbounded may well arise. This derivative may often be related to consumption of a certain resource, such as, fuel, finances, nutritious substance, battery charge, and so on–in engineering, in economics, in biology, and so forth. Then, in order to minimize the consumption of this resource x, the cost functional

$$x(t_1) - x(t_0) = \int_{t_0}^{t_1} \dot{x} dt \to \min$$

is considered, which, after the extension procedure, leads to the integral functional

$$\int_{t_0}^{t_1} d\nu \to \min .$$

The scalar measure ν expresses the above-discussed consumption rate. Consumption rate can take very large values at some instances of time—notably, when the impulse occurs—in which case, this measure is supposed to have atoms at such points of time. This integral form of the cost is, of course, already included in (1.6).

Let us proceed with a strict mathematical theory for this class of impulsive control problems. The triple (x, u, μ) is called *impulsive control process* (or, for short, control process), if (1.5) holds. A control process is said to be admissible, if it satisfies all the constraints of problem (1.4), that is, $p = (x(t_0), x(t_1)) \in S$, $u(t) \in U$ a.e., $\mu \geq 0$. An admissible process $(\hat{x}, \hat{u}, \hat{\mu})$ is said to be optimal if, for any admissible process (x, u, μ), the inequality $\varphi(\hat{p}) \leq \varphi(p)$ is satisfied, where $\hat{p} = (\hat{x}(t_0), \hat{x}(t_1))$.

Next, we present sufficient conditions for the existence of a solution and derive necessary conditions for optimality in the form of the Pontryagin maximum principle [8] for Problem (1.4), which are presented in Theorem 1.1 and Theorem 1.2, respectively.

The following basic notations are adopted throughout the book. Let μ be a Borel vector-valued measure. Then,

$$F(t; \mu) := \int_{[t_0,t]} d\mu, \quad t > t_0, \quad F(t_0; \mu) := 0,$$

$$\mathrm{Ds}(\mu) := \{\tau \in T : \mu(\{\tau\}) \neq 0\},$$

$$\mathrm{Cs}(\mu) := T \setminus \mathrm{Ds}(\mu) \cup \{t_0\} \cup \{t_1\}.$$

Thus, $\mathrm{Ds}(\mu)$ is the set of points of time in which discontinuities of function $F(\cdot; \mu)$ occur, while $\mathrm{Cs}(\mu)$ is the set of its continuity points plus the endpoints.

Recall that a sequence of vector-valued Borel measures μ_i on T converges weakly-* to μ, denoted as $\mu_i \overset{w*}{\to} \mu$, if

$$\lim_{i\to\infty} \int_T \langle \phi(t), d\mu_i \rangle = \int_T \langle \phi(t), d\mu \rangle$$

for any continuous function $\phi : T \to \mathbb{R}^k$. This is equivalent to the componentwise weak-* convergence.

1.3 Existence Theorem

First, we observe the fact that linear problems always have a solution as soon as there exists at least one admissible process, and the impulsive control μ is bounded w.r.t. the total variation norm $\|\mu\|$.

Theorem 1.1 *Let there exist an admissible process $(\bar{x}, \bar{u}, \bar{\mu})$ for Problem (1.4). Suppose that the set S is compact, and there exists a constant $\kappa > 0$ such that $\|\mu\| \leq \kappa$ for any admissible process (x, u, μ) for which $\varphi(p) \leq \varphi(\bar{p})$, where $\bar{p} = (\bar{x}(t_0), \bar{x}(t_1))$. Then, there exists an optimal process.*

Proof It is not restrictive to consider that the estimate $\|\mu\| \leq \kappa$ holds over all admissible processes (x, u, μ). Indeed, this conclusion is simple to reach by imposing the additional endpoint constraint $\varphi(p) \leq \varphi(\bar{p})$.

First, let us note that the set of admissible trajectories (which is not empty by assumption) is bounded in the $\|\cdot\|_{L_\infty}$ norm. Indeed, let $x(t)$ be any admissible trajectory of Problem (1.4). Then, from the Gronwall inequality (Exercise 1.2) and from (1.5), since U is bounded, we have that

$$|x(t)| \leq (|x_0| + \|B\|_{L_\infty} \cdot \|u\|_{L_1} + \|f\|_{L_1} + \|C\|_C \cdot \|\mu\|) \cdot e^{\|A\|_{L_\infty} \cdot (t_1 - t_0)} \leq \text{const} \tag{1.7}$$

for all $t \in T$, while the constant in (1.7) does not depend on $x(\cdot)$.

In view of this estimate and from the inequality of Exercise 1.3, we conclude that

$$\operatorname{Var} x|_T \leq \text{const}. \tag{1.8}$$

Now, consider a minimizing sequence (x_i, u_i, μ_i) to Problem (1.4), that is, such a sequence that

$$\varphi(p_i) \to \inf_{p=(x(t_0),x(t_1))} \varphi(p), \tag{1.9}$$

where $p_i = (x_i(t_0), x_i(t_1))$ and the infimum is taken over all admissible trajectories $x(t)$. Note that $\|\mu_i\| \leq \kappa \ \forall i$. In view of (1.7), (1.8), and since the set S is compact, by the Helly selection theorem, [3], a subsequence can be extracted ensuring that $x_i(t) \to x(t)$ for a.a. t, $\mu_i \overset{w*}{\to} \mu$, and $p_i \to p = (x(t_0), x(t_1))$ as $i \to \infty$, where $x \in BV(T; \mathbb{R}^n)$ is a right-continuous on (t_0, t_1) vector-valued function of bounded variation, and $\mu \in C(T; \mathbb{R}^k)$ is a vector-valued measure. By virtue of the weak-* convergence of measures, the condition $\mu_i \geq 0$ implies that $\mu \geq 0$. In view of the boundedness of U, one has the estimate $\|u_i\|_{L_2} \leq \text{const} \ \forall i$. Then, by using the weak sequential compactness of the unit ball in $L_2(T; \mathbb{R}^m)$, we may also consider, after extracting a subsequence, that $u_i \overset{w}{\to} u$ for some $u \in L_2(T; \mathbb{R}^m)$.

Since each term of the sequence $\{x_i\}$ is defined by

$$x_i(t) = x_0 + \int_{t_0}^{t} \Big(A(s)x_i(s) + B(s)u_i(s) + f(s)\Big)ds + \int_{[t_0,t]} C(s)d\mu_i, \quad \forall t > t_0,$$

then, by passing to the limit for a.a. t such that $x_i(t) \to x(t)$ and by using the fact that function $x(\cdot)$ is right-continuous on (t_0, t_1), we obtain

$$x(t) = x_0 + \int_{t_0}^{t} \Big(A(s)x(s) + B(s)u(s) + f(s)\Big)ds + \int_{[t_0,t]} C(s)d\mu, \quad \forall t > t_0.$$

So, (x, u, μ) is a control process. Let us show that it is admissible. Since the set U is convex and closed, the set of functions $\mathscr{U} = \{u \in \mathrm{L}_2(T; \mathbb{R}^m) : u(t) \in U \text{ a.a. } t\}$ is also convex and closed. (Use Exercise 1.4 to ensure the closedness.) By construction, $u_i \xrightarrow{w} u$. Then, by virtue of the Mazur lemma, [2], for all positive integer N, there exists a function $v_N \in \mathrm{L}_2(T; \mathbb{R}^m)$ which is a convex combination of $u_1, u_2, ..., u_N$, such that $v_N \to u$ as $N \to \infty$ strongly in L_2. Then, we have the following chain of implications: $u_i \in \mathscr{U} \Rightarrow v_N \in \mathscr{U} \Rightarrow u \in \mathscr{U} \Rightarrow u(t) \in U$ for a.a. $t \in T$.

Since $p \in S$, it has just been shown that the process (x, u, μ) is admissible. However, $\varphi(p_i) \to \varphi(p)$, so the infimum in (1.9) is attained at $p = (x(t_0), x(t_1))$, and then, the process (x, u, μ) is optimal. □

Remark 1.1 The condition that $\|\mu\| \le \kappa \; \forall \, (x, u, \mu)$ such that $\varphi(p) \le \varphi(\bar{p})$ is crucial for the existence of a solution. Without it, Theorem 1.1 is a false assertion even if the infimum in Problem (1.4) is finite. If we add to the problem formulation the constraint $\|\mu\| \le \kappa$, then this condition will be a priori satisfied. At the same time, the constraint $\|\mu\| \le \kappa$ can always be added to the statement of Problem (1.4) in an obvious way: For that, we need to introduce new state variables y^j such that $y^j(t_0) = 0$, $y^j(t) = \int_{[t_0,t]} d\mu^j$, $t > t_0$, $j = 1, \ldots, k$ and to consider in the original problem an additional endpoint constraint $\sum_j y^j(t_1) \le \kappa$.

Remark 1.2 The constraint $\|\mu\| \le \kappa$ is obviously not restrictive if the minimizing functional is given by the formula $\|\mu\| = \sum_j \int_{[t_0,t_1]} d\mu^j$. This is the case in many actual applications, for example, in optimizing consumption of resources, such as fuel in spacecraft maneuvering problems (see Sect. 1.1).

Remark 1.3 Instead of the functional $\varphi(p)$ in Problem (1.4), consider a more general functional

$$\varphi(p) + \int_{t_0}^{t_1} \xi(x, t)dt,$$

where ξ is a continuous scalar function. Then, Theorem 1.1 is still valid. This easily follows from its proof.

1.4 Maximum Principle

Let us proceed with the necessary optimality conditions for Problem (1.4). Denote by H, the classical Hamilton–Pontryagin function

$$H(x, u, \psi, t) := \langle A(t)x + B(t)u + f(t), \psi \rangle,$$

where $\psi \in \mathbb{R}^n$ is the so-called co-state, or adjoint, variable. Denote by Q, the following linear vector-valued function

$$Q(t, \psi) := C(t)^* \psi.$$

Denote by $N_S(p)$ the limiting normal cone to the set S at point p in the sense proposed by Mordukhovich [6]. For $p \in S$, this cone is defined by the following relation:

$$N_S(p) := \operatorname*{Limsup}_{y \to p} \operatorname{cone}(y - \Pi_S(y)).$$

Here, $\Pi_S(y)$ is the Euclidean projection of vector y on the set S, i.e.

$$\Pi_S(y) = \{p \in S : \ |y - p| = \operatorname{dist}(y, S)\},$$

where $\operatorname{dist}(y, S) = \inf_{p \in S} |y - p|$ is the distance from the point to the set; cone stands for the conic hull of a set, and Limsup stands for the sequential Painlevé–Kuratowski upper/outer limit of a family of sets; see [7].

Theorem 1.2 *Let $(\hat{x}, \hat{u}, \hat{\mu})$ be an optimal process in Problem (1.4). Then, there exist a number $\lambda \geq 0$ and an absolutely continuous vector-valued function $\psi : T \to \mathbb{R}^n$ such that*

$$\lambda + |\psi(t)| \neq 0 \ \ \forall t \in T, \tag{1.10}$$

$$\dot{\psi}(t) = -\frac{\partial H}{\partial x}(t) = -A(t)^* \psi(t), \tag{1.11}$$

$$(\psi(t_0), -\psi(t_1)) \in \lambda \varphi'(\hat{p}) + N_S(\hat{p}), \tag{1.12}$$

$$\max_{u \in U} H(u, t) = H(t) \ \ a.e. \tag{1.13}$$

$$Q(t) \leq 0 \ \ \forall t \in T, \tag{1.14}$$

$$Q^j(t) = 0 \ \ \forall t \in \operatorname{supp} \hat{\mu}^j, \ \ j = 1, .., k. \tag{1.15}$$

Here, $\hat{p} = (\hat{x}(t_0), \hat{x}(t_1))$.

Here, and from now on, the following convention is adopted. If some of the arguments of H, Q, or of their partial derivatives are omitted, then it means that the extremal values substitute the missing arguments everywhere, i.e., $H(t) = H(\hat{x}(t), \hat{u}(t), \psi(t), t)$, $H(u, t) = H(\hat{x}(t), u, \psi(t), t)$.

Before the proof, let us comment on the comparison between the maximum principles for the impulsive and for the conventional optimal control problems. Conditions (1.10)–(1.13) are just the same as those in the conventional maximum principle, as in [8]: It is necessary to maximize the function $\langle B^*\psi(t), u\rangle$ over set U, while its maximum is attained at $u = \hat{u}(t)$ for a.a. t. At the same time, Conditions (1.14), (1.15) represent the maximum condition for the impulsive part of the dynamics. Indeed, these conditions state that, for any vector-valued Borel measure $\mu \geq 0$, it holds that

$$\int_B \langle Q(t), d\mu\rangle \leq 0,$$

where $B \subseteq T$ is an arbitrary Borel subset. However,

$$\int_B \langle Q(t), d\hat{\mu}\rangle = 0 \;\; \forall\, B.$$

Thus, the maximum is attained at the optimal measure $\hat{\mu}$. In many instances, the optimal measure is likely discrete, that is, $\hat{\mu}^j = \sum_r \delta_{\tau_r^j}$, i.e., it is a finite sum of Dirac's measures. Then, according to the above-stated maximum principle, in order to find the moments of optimal impacts τ_r^j, it is necessary to solve the equation

$$\langle c^j(\tau), \psi(\tau)\rangle = 0,$$

where $c^j(\cdot)$ is the jth column of the matrix C, while the function $\psi(t)$ satisfies (1.10)–(1.13).

Proof Let us use the so-called penalty function method; see, for example, in [1]. Let the process $(\hat{x}, \hat{u}, \hat{\mu})$ be optimal. It is not restrictive to consider that $\varphi(\hat{p}) = 0$, where $\hat{p} = (\hat{x}(t_0), \hat{x}(t_1))$. There exists a sequence of continuous controls $\bar{u}_i \in \mathrm{C}(T; \mathbb{R}^m)$ such that $\bar{u}_i(t) \to \hat{u}(t)$ a.e. (see [3]). Let us set

$$\bar{y}_i(t) := \int_{t_0}^{t} \bar{u}_i(s)ds.$$

In view of Exercise 1.5, there exists a sequence of absolutely continuous measures $\bar{\mu}_i \xrightarrow{w*} \hat{\mu}$ such that $d\bar{\mu}_i = \bar{m}_i(t)dt$, where $\bar{m}_i(t) \in \mathrm{C}(T; \mathbb{R}^k)$. Let $\bar{x}_i$ be the trajectory corresponding to the triple $(\hat{x}_0, \bar{u}_i, \bar{\mu}_i)$ w.r.t. (1.5).

Let us denote $K_i := \|\bar{m}_i\|_{\mathrm{C}}$, $\bar{\varepsilon}_i := \varphi(\bar{p}_i)$, $A_i = (\mathrm{dist}(\bar{p}_i, S))^{-1/2}$, where $\bar{p}_i = (\bar{x}_i(t_0), \bar{x}_i(t_1))$. By Exercise 1.6, we have $\bar{x}_i(t) \to \hat{x}(t)$ $\forall\, t \in \mathrm{Cs}(\hat{\mu})$, and, then, $\bar{\varepsilon}_i \to 0$, $A_i \to \infty$ as $i \to \infty$.

For every positive integer i, consider the auxiliary i-problem:

$$\begin{aligned}
\text{Minimize}\ \ & J_i(\wp) = \varphi(p) + A_i \operatorname{dist}(p, S) + |p - \hat{p}|^2 + |z_1 - F(t_1; \hat{\mu})|^2 + \\
& + \int_{t_0}^{t_1} \left(|y - \bar{y}_i(t)|^2 + |z - F(t; \bar{\mu}_i)|^2 \right) dt \\
\text{subject to:}\ \ & \dot{x} = A(t)x + B(t)u + f(t) + C(t)m, \\
& \dot{y} = u, \\
& \dot{z} = m, \\
& p = (x_0, x_1),\ |p - \hat{p}| \le 1, \\
& y_0 = 0,\ z_0 = 0,\ |z_1| \le \|\hat{\mu}\| + 1, \\
& u \in U,\ m \in M_i,
\end{aligned} \tag{1.16}$$

where $M_i := \{y \in \mathbb{R}^k : y^j \in [0, K_i]\}$, by $\wp$, the triple (x, u, m) is denoted, and by $m \in \mathbb{R}^k$–the usual bounded control parameter which replaces the measure μ.

Note that, for sufficiently large i, the process $\bar{\wp}_i = (\bar{x}_i, \hat{u}, \bar{m}_i)$ in Problem (1.16) is admissible. Then, by Theorem 1.1, and in view of Remark 1.3, this problem has a solution. Let us denote it by $\wp_i = (x_i, u_i, m_i)$. Let $\mu_i \in C^*(T; \mathbb{R}^k)$ be such a measure that $d\mu_i = m_i(t)dt$. Since

$$|p_i| + \|\mu_i\| + \|u_i\|_{L_2} \le \text{const},$$

by compactness arguments, we have that, after extracting a subsequence, $p_i \to p$, $u_i \xrightarrow{w} u \in L_2(T; \mathbb{R}^m)$, and $\mu_i \xrightarrow{w*} \mu$ for some vector-valued measure $\mu \in C^*(T; \mathbb{R}^k)$ such that $\mu \ge 0$.

Let $x(t)$ be the trajectory from (1.5) corresponding to the obtained x_0, u, μ. By Exercise 1.6, we know that $p = (x(t_0), x(t_1))$. Let us show that $p = \hat{p}$, $u = \hat{u}$, and $\mu = \hat{\mu}$.

Denote $y_i(t) = \int_{t_0}^{t} u_i(s)ds$, $z_i(t) = \int_{t_0}^{t} m_i(s)ds$. In view of Exercises 1.6 and 1.7,

$$y_i(t) \rightrightarrows y(t) = \int_{t_0}^{t} u(s)ds, \quad z_i(t) \to z(t) = F(t; \mu)\ \ \forall t \in \mathrm{Cs}(\mu),$$

$$\bar{y}_i(t) \rightrightarrows \hat{y}(t) = \int_{t_0}^{t} \hat{u}(s)ds, \quad i \to \infty.$$

It is clear that

$$J_i(\wp_i) \le J_i(\bar{\wp}_i) =$$

$$= A_i \operatorname{dist}(\bar{p}_i, S) + |\bar{p}_i - \hat{p}|^2 + \varphi(\bar{p}_i) = A_i^{-1} + |\bar{p}_i - \hat{p}|^2 + \bar{\varepsilon}_i \to 0.$$

Therefore, due to the fact that the function $\varphi(p)$ is bounded from below in (1.16), we deduce

$$A_i \operatorname{dist}(p_i, S) \le \text{const} \;\Rightarrow\; \operatorname{dist}(p_i, S) \to 0 \;\Rightarrow\; p \in S.$$

By applying the same arguments as in Theorem 1.1, and by using Mazur's lemma, [2], we obtain that $u(t) \in U$ a.e. Hence, the process (x, u, μ) is admissible in Problem (1.4). Then, $0 = \varphi(\hat{p}) \le \varphi(p)$. Therefore, we have

$$0 \le \varphi(p) + |p - \hat{p}|^2 + |z_1 - F(t_1; \hat{\mu})| +$$

$$+ \int_{t_0}^{t_1} \Big(|y(t) - \hat{y}(t)|^2 + |z(t) - F(t; \hat{\mu})|^2 \Big) dt =$$

$$= \lim_{i\to\infty} J_i(\wp_i) \le \lim_{i\to\infty} J_i(\hat{\wp}_i) = 0.$$

Then, $p = \hat{p},\ u = \hat{u}$, and $\mu = \hat{\mu}$.

In view of Exercise 1.6, we obtain $x_i(t) \to \hat{x}(t)\ \forall\, t \in \mathrm{Cs}(\hat{\mu})$.

Note that Problem (1.16) is a conventional optimal control problem and it is unconstrained when number i is sufficiently large. Let us apply the nonsmooth version of the maximum principle from [6] to Problem (1.16). Assume that i is sufficiently large. For each such i, there exist a number $\lambda_i > 0$, and absolutely continuous functions ψ_i, ϕ_i, ξ_i, such that

$$\lambda_i + \|\psi_i\|_C = 1, \tag{1.17}$$

$$\begin{cases} \dot{\psi}_i = -A^*(t)\psi_i, \\ \dot{\phi}_i = 2\lambda_i(y_i - \bar{y}_i(t)), \\ \dot{\xi}_i = 2\lambda_i(z_i - F(t; \bar{\mu}_i)), \end{cases} \tag{1.18}$$

$$\begin{cases} (\psi_i(t_0), -\psi_i(t_1)) \in \lambda_i \varphi'(p_i) + \lambda_i A_i \partial \operatorname{dist}(p_i, S) + 2\lambda_i(p_i - \hat{p}), \\ \phi_i(t_1) = 0, \\ \xi_i(t_1) = -2\lambda_i(z_i(t_1) - F(t_1, \hat{\mu})), \end{cases} \tag{1.19}$$

$$\max_{u\in U} \big(H_i(u, t) + \langle \phi_i(t), u \rangle \big) = H_i(t) + \langle \phi_i(t), u_i(t) \rangle \ \ a.e., \tag{1.20}$$

$$\max_{m\in M_i} \big(\langle Q_i(t), m \rangle + \langle \xi_i(t), m \rangle \big) = \langle Q_i(t), m_i(t) \rangle + \langle \xi_i(t), m_i(t) \rangle \ \ a.e. \tag{1.21}$$

Above, if some function of x, u, ψ has the subscript i and, at the same time, some of its arguments are omitted, then this means that the extremal values $x_i(t), u_i(t), \psi_i(t)$ replace the omitted arguments. For example, $H_i(t) = H(x_i(t), u_i(t), \psi_i(t), t)$.

In view of (1.17), by Arzela–Ascoli theorem, we have, after extracting a subsequence, that $\lambda_i \to \lambda$, $\psi_i \rightrightarrows \psi$ as $i \to \infty$. Let us show that λ, ψ satisfy conditions (1.10)–(1.15).

First, let us note that $\phi_i \rightrightarrows 0$ and $\xi_i \rightrightarrows 0$ as $i \to \infty$. Indeed, as it was shown above $y_i - \bar{y}_i \rightrightarrows 0$, $z_i(t) - F(t; \bar{\mu}_i) \to 0$ a.a. t. Then, it follows from (1.17), (1.18) and (1.19) that $\phi_i \rightrightarrows 0$, $\xi_i \rightrightarrows 0$.

From (1.18), it follows that

$$\lambda + \|\psi\|_C = 1. \tag{1.22}$$

After passing to the limit in (1.18), we obtain (1.11). Then, Condition (1.10) is a simple corollary of (1.22) and the Cauchy theorem.

By using the properties of subdifferential of the distance function (see [7], Sect. 1.3.3, Theorems 1.97 and 1.105 therein), the upper semicontinuity of the cone $N_S(p)$, from (1.19), we derive (1.12).

Let $\mathscr{U} = \{u \in L_2(T; \mathbb{R}^m) : u(t) \in U \text{ a.a. } t\}$. By integrating (1.20), using that function H is affine w.r.t. u, and in view of the weak convergence, after passing to the limit, we obtain the relation

$$\max_{u(\cdot)\in\mathscr{U}} \int_{t_0}^{t_1} H(u,t)dt = \int_{t_0}^{t_1} H(t)dt,$$

which is equivalent to (1.13) by virtue of Exercise 1.8.

Let us show that (1.14) is true. Indeed, $Q_i(t) \rightrightarrows Q(t)$. However, $\xi_i(t) \rightrightarrows 0$. So, by assuming for some j that $Q^j(t) > 0$ on some interval $I = [a, b] \subseteq T$, in view of (1.21), it follows that, for large i, $m_i^j(t) = K_i$ for a.a. $t \in I$. However, $K_i \to \infty$. Therefore, $\|\mu_i^j\| \to \infty$ which is not possible. Thus, (1.14) is true.

From (1.21),

$$\langle Q_i(t), m_i(t)\rangle + \langle \xi_i(t), m_i(t)\rangle \geq 0 \quad \text{a.a. } t \in T.$$

By integrating this inequality, and by passing to the limit as $i \to \infty$,

$$\int_T \langle Q(t), d\hat{\mu}\rangle \geq 0.$$

From here, and from (1.14), it follows (1.15). □

Remark 1.4 Above, linear optimal impulsive control problems with fixed time were considered. Nevertheless, an important class of linear problems is presented by time-optimal control problems. This purports the minimization in (1.4), instead of $\varphi(p)$, the time $t_1 - t_0$ which the material point $x(t)$ takes to move from x_0 to x_1. Then, time interval T is not assumed as fixed. The maximum principle for time-optimal control problems also includes the so-called time transversality conditions. The case of impulsive control problems with nonfixed time will be considered in Chap. 4.

1.5 Exercises

Exercise 1.1 Construct the extensions for problems (1.1)–(1.3) raised in Examples 1.1–1.3. Find the extremals by applying the maximum principle of Theorem 1.2. Also find, numerically, the widths of steps $t_{0,n}$, $t_{1,n}$ in Example 1.3 and verify their convergence to zero.

Exercise 1.2 Prove the Gronwall inequality. That is, for a given scalar nonnegative function $x \in \mathrm{C}(T; \mathbb{R})$ such that

$$x(t) \le a + \int_{t_0}^{t} b \cdot x(\varsigma) d\varsigma,$$

where $a, b \ge 0$, prove that

$$x(t) \le a \cdot e^{b \cdot (t - t_0)}.$$

Exercise 1.3 Given $x(t) = \int_{[t_0, t]} f(s)\, d\mu$, $t > t_0$, with $x(t_0) = 0$, where $f \in \mathrm{C}(T; \mathbb{R})$, $\mu \in \mathrm{C}^*(T; \mathbb{R})$. In this case, we say that the function x is generated by the measure μ. Prove the following inequality for the generated function x

$$\sup_{s \in [a,b]} |x(s) - x(a)| \le \mathrm{Var}\,|_a^b[x] \le \max_{s \in [a,b]} |f(s)| \times |\mu|([a, b]) \;\; \forall a, b \in T.$$

Exercise 1.4 Let $u_i \to u$ in $\mathrm{L}_2(T; \mathbb{R}^m)$. Then, there exists a subsequence u_{i_k} which converges to u almost everywhere.

Exercise 1.5 The subset $\mathscr{A} \subseteq \mathrm{C}^*(T; \mathbb{R}^k)$ of absolutely continuous measures μ such that $\frac{d\mu}{dt} = m(t)$, where $m \in \mathrm{C}(T; \mathbb{R}^k)$, is weakly-* dense in $\mathrm{C}^*(T; \mathbb{R}^k)$. That is, for every $\mu \in \mathrm{C}^*(T; \mathbb{R}^k)$, there exists a sequence of measures $\mu_i \in \mathscr{A}$ such that $\mu_i \overset{w*}{\to} \mu$.

Exercise 1.6 Let the triples (x_i, u_i, μ_i) be control processes in Problem (1.4). Let $u_i \overset{w}{\to} u \in \mathrm{L}_2(T; \mathbb{R}^m)$, $\mu_i \overset{w*}{\to} \mu \in \mathrm{C}^*(T; \mathbb{R}^k)$, and $x_i(t_0) \to x_0$. Then,

$$x_i(t) \to x(t) \;\; \forall t \in \mathrm{Cs}(\mu),$$

where $x(t)$ is the existing and unique solution to (1.5) w.r.t. x_0, u, μ.

Exercise 1.7 Let $u_i \xrightarrow{w} u \in \mathrm{L}_2(T;\mathbb{R}^m)$, and $\|u_i\|_{\mathrm{L}_\infty} \leq \text{const}\ \forall i$. Then,

$$\int_{t_0}^{t} u_i(s)ds \rightrightarrows \int_{t_0}^{t} u(s)ds.$$

Exercise 1.8 Let $h(u,t)$ be a scalar bounded function continuous w.r.t. u and measurable w.r.t. t. Consider the set $\mathscr{U} = \{u \in \mathrm{L}_2(T;\mathbb{R}^m) : u(t) \in U \text{ a.a. } t\}$. Show that the two maximum conditions are equivalent:

$$\max_{u\in U} h(u,t) = h(\bar{u}(t),t) \ \text{ a.a. } t \quad \Leftrightarrow$$

$$\max_{u(\cdot)\in\mathscr{U}} \int_{t_0}^{t_1} h(u(t),t)dt = \int_{t_0}^{t_1} h(\bar{u}(t),t)dt,$$

where $\bar{u} \in \mathscr{U}$.

Exercise 1.9 Generalize Remark 1.3: The assertion of Theorem 1.1 is still valid if the cost functional is replaced by

$$\varphi(p) + \int_{t_0}^{t_1} \xi(x,u,t)dt,$$

where ξ is a continuous scalar function which is convex w.r.t. u for each x, t.

Exercise 1.10* Let the maps A, B, C and f be smooth. Denote

$$h(t) := \max_{u\in U} H(u,t), \quad t \in T,$$

The maximum principle can be supplemented by the following condition:

$$h(t) = \kappa + \int_{t_0}^{t} \frac{\partial H}{\partial t}(\tau)d\tau + \int_{[t_0,t]} \frac{\partial Q}{\partial t}(\tau)d\hat{\mu} \ \ \forall t \in (t_0,t_1], \ \ h(t_0) = \kappa, \qquad (1.23)$$

where $\kappa \in \mathbb{R}$ is some constant. Prove this by using merely the penalty function method. Develop the idea of smooth approximations in the proof of Theorem 1.2.

Exercise 1.11* Let $C(\cdot) = 0$. Then, as the classic theory states, the Formula (1.23) is merely the consequence of extremality. (Optimality is redundant.) So, (1.23) follows from (1.11)–(1.15). Prove the same in the impulsive control context, that is, when $C(\cdot) \neq 0$.

References

1. Arutyunov, A.V.: Optimality Conditions. Abnormal and degenerate problems. Mathematics and its applications. Kluwer Academic Publishers, Dordrecht (2000)
2. Ioffe, A., Tikhomirov, V.: Studies in Mathematics and Its Applications, vol. 6. Elsevier Science, North-Holland, Amsterdam (1979)
3. Kolmogorov, A., Fomin, S.: Introductory Real Analysis (Dover Books on Mathematics). Dover Publications (1975)
4. Lawden, D.: Optimal Trajectories for Space Navigation (Butterworths mathematical texts) (1963)
5. Lawden, D.: Rocket trajectory optimization—1950–1963. J. Guid. Control Dyn. **14**(4), 705–711 (1991)
6. Mordukhovich, B.: Maximum principle in the problem of time optimal response with nonsmooth constraints. J. Appl. Math. Mech. **40**, 960–969 (1976)
7. Mordukhovich, B.: Variational Analysis and Generalized Differentiation I. Basic Theory, Grundlehren der Mathematischen Wissenschaften [Fundamental Principles of Mathematical Sciences], vol. 330. Springer, Berlin (2006)
8. Pontryagin, L., Boltyanskii, V., Gamkrelidze, R., Mishchenko, E.: Mathematical theory of optimal processes. Transl. from the Russian ed. by L.W. Neustadt, First Edition. Interscience Publishers, Wiley Inc. (1962)

Chapter 2
Impulsive Control Problems Under Borel Measurability

Abstract In this chapter, the complexity of the dynamical control system in the optimal control problem under extension increases. Herein, it is not linear w.r.t. x and u but is still linear w.r.t. the impulsive control variable. Moreover, the matrix-multiplier for the impulsive control depends on the conventional control $u(\cdot)$ given by Borel functions. The right-hand side of the dynamical system is assumed to be Borel w.r.t. u. The results of the first chapter are derived for this more general formulation. The concept of extension itself does not change so far, as the space of Borel measures yet suffices to describe all feasible trajectories. The chapter ends with seven exercises.

2.1 Introduction

In this chapter, impulsive control problem formulation involves some extra complexity relative to the one considered in the previous chapter. More specifically, the case is studied when the matrix G multiplying the control measure μ depends, not only on the time variable t, but also on the control variable u, that is, when $G = G(u, t)$. Moreover, the conventional component f of the impulsive control dynamics is now nonlinear, while the right-hand side (f, G) is assumed to be merely Borel measurable w.r.t. u. This case is referred to as impulsive control problems under the condition of Borel measurability. Some results in the previous chapter are generalized, while the maximum principle is proved under rather weak assumptions involving nonsmooth aspects. A specific line of proof allows us to overcome the difficulties arising from the condition of Borel measurability. In the remaining chapters, the Borel measurability is replaced with standard hypothesis of continuity w.r.t. u.

The control function $u(\cdot)$ is also assumed to be merely Borel measurable. The following observation should be stressed. In what concerns the class of controls $u(\cdot)$, it is not really significant, w.r.t. the control systems studied in this book, whether $u(\cdot)$ is Lebesgue measurable,[1] or only Borel measurable. Indeed, as follows from measure theory, see Exercise 2.1, by changing values of a Lebesgue measurable control $u(\cdot)$ on some zero measure set, it is possible to make this control Borel

[1]That is, measurable w.r.t. the Lebesgue measure ℓ generated by length and w.r.t. the Lebesgue–Stieltjes measure μ, which is the unique Lebesgue completion of the corresponding Borel measure.

© Springer Nature Switzerland AG 2019
A. Arutyunov et al., *Optimal Impulsive Control*, Lecture Notes in Control and Information Sciences 477, https://doi.org/10.1007/978-3-030-02260-0_2

measurable. However, zero measure sets do not affect the evolution of the trajectory $x(\cdot)$. Therefore, the case of Lebesgue measurable controls can easily be reduced to the case of Borel measurable controls.

On the one hand, this means that the generality, or the added benefits, of using Lebesgue controls rather than Borel controls is not substantive in the context of the optimal control theory requirements as a whole. On the other hand, the classical approach adopts the more general Lebesgue concept due to a certain convenience arising from useful properties of the Lebesgue measure, notably its completeness. Throughout this book, the line of exposition based on the Lebesgue controls is adhered to, with the exception of the current chapter where it can be seen that it is not restrictive to consider the controls measurable merely in the Borel sense. This chapter also invokes some elements of the nonsmooth analysis, [1, 4, 6, 7], as the data are supposed to be Lipschitz continuous w.r.t. the state variable. The results and methods developed in the other chapters for the smooth data w.r.t. x can be readily modified and applied to the case when the data are Lipschitz continuous w.r.t. x, as in this chapter.

The solution concept used in this chapter corresponds to the classical solution which is based on the standard integration. In fact, it is the same as in the previous chapter. However, in view of the fact that G now depends on u, this solution does not suffice to ensure the well posedness w.r.t. approximations in the weak-* topology due to possible discontinuities of the control function u, see Exercise 2.2. A full proper, or well posed, extension of the case $G = G(u,t)$ will be presented later in Chap. 6.

The proof for the main result of this chapter (maximum principle) is different from the one used in the previous chapter. It relies on the Ekeland variational principle and a nonsmooth approach of boundary control process, in which the maximum principle for the given impulsive control problem is extracted from the necessary conditions for a certain related Ψ-boundary control process. The material of this chapter and the idea behind the proof are based on the work [8].

2.2 Problem Statement

Consider the following optimal impulsive control problem

$$\begin{aligned}
&\text{Minimize } \varphi(p)\\
&\text{subject to } dx = f(x,u,t)dt + G(u,t)d\mu, \quad t \in T,\\
&\qquad p = (x_0, x_1) \in S,\\
&\qquad u(t) \in U \quad (\ell + |\mu|)\text{-a.e.},\\
&\qquad \text{range}(\mu) \subset K,
\end{aligned} \tag{2.1}$$

where $T = [t_0, t_1]$ is the time interval, $x_0 = x(t_0)$, $x_1 = x(t_1)$ are the endpoints, $f : \mathbb{R}^n \times \mathbb{R}^m \times T \to \mathbb{R}^n$, $G : \mathbb{R}^m \times T \to \mathbb{R}^{n\times k}$ are given maps, S is a closed subset of $\mathbb{R}^{2n}$, U is a Borel subset of $\mathbb{R}^m$, and K is a closed convex cone in $\mathbb{R}^k$.

The control policy is a pair (u, μ), where $u(\cdot)$ is a Borel function defined over T with values in U, and μ is a Borel measure on T with values in K. This means that $u(t) \in U$ for almost all t w.r.t. the Lebesgue ℓ and the total variation $|\mu|$ measures and that $\mu(B) \in K \ \forall B \in \mathscr{B}$, where $\mathscr{B} = \mathscr{B}(T)$ stands for the σ-algebra of Borel subsets of T. A trajectory $x(\cdot) : T \to \mathbb{R}^n$, associated with control policy (u, μ), is a function of bounded variation such that

$$x(t) = x_0 + \int_{t_0}^{t} f(x(s), u(s), s)ds + \int_{[t_0,t]} G(u(s), s)d\mu, \quad t \in (t_0, t_1], \quad x(t_0) = x_0,$$

where both integrals are supposed to exist.[2] If such a function $x(\cdot)$ exists for some x_0, then the triple (x, u, μ) is called control process. The control process (x, u, μ) is called feasible if all the constraints imposed in Problem (2.1) are satisfied. The control process $(\hat{x}, \hat{u}, \hat{\mu})$ is called optimal or solution to (2.1) if $\varphi(\hat{p}) \le \varphi(p)$ for all feasible (x, u, μ), where $\hat{p} = (\hat{x}_0, \hat{x}_1)$, $p = (x_0, x_1)$.

Similarly to that which was considered in Chap. 1, Problem (2.1) may be regarded as an extension of a conventional optimal control problem, in which some control components, appearing affinely, in the dynamics may take unbounded values along certain directions, that is,

$$\begin{aligned}
&\text{Minimize } \varphi(p) \\
&\text{subject to } \dot{x} = f(x, u, t) + G(u, t)v, \quad t \in T, \\
&\qquad p = (x_0, x_1) \in S, \\
&\qquad u(t) \in U, \ \ v(t) \in K \text{ a.a. } t.
\end{aligned}$$

The class of control processes for this problem can be related to the ones for (2.1) via the mapping $(x, u, v) \to (x, u, \mu)$, where the measure μ is given by $\mu(A) = \int_A v(t)dt$, for any $A \in \mathscr{B}$. However, unlike the extension in Chap. 1, this one is not well posed w.r.t. the weak-* topology of the space of measures, see Eq. 2.1. This is due to the fact that matrix G depends on control variable u. Thus, a more delicate topology is required to ensure the well posedness of the extension.[3]

Let us recall some standard concepts and notation.

The total variation measure of μ, appearing above and denoted by $|\mu|$, is given by $|\mu| := \sum_{i=1}^{k} \mu_i^+ + \mu_i^-$ where $\mu_i = \mu_i^+ - \mu_i^-$ is the Jordan decomposition of the ith component of the measure μ.

The support function $\sigma_B(y)$ of the set $B \subset \mathbb{R}^n$ at the point $y \in \mathbb{R}^n$ is given by

$$\sigma_B(y) := \sup_{x \in B} \langle y, x \rangle.$$

---

[2] While the first integral by ds is understood in the conventional Lebesgue sense, the second integral by the vector-valued Borel measure $d\mu$ is understood in the sense of the Lebesgue–Stieltjes integral of the vector-valued Borel function $G(u(s), s)w(s)$ by the Borel measure $|\mu|$, where function $w(\cdot)$ means the Radon–Nikodym derivative of μ w.r.t. $|\mu|$.

[3] This case is analyzed in Chap. 6; see formula (6.4) and also Exercise 6.10.

For a given closed set S and a point $p \in S$, as usual, the limiting normal cone at point p is denoted by $N_S(p)$, and, for a Lipschitz continuous function φ, the limiting subdifferential at p is denoted by $\partial\varphi(p)$. While the notion of the limiting normal cone was given in the previous chapter, the notion of the limiting subdifferential is defined via the normal cone as follows, [4]:

$$\partial\varphi(p) := \{y \in \mathbb{R}^{2n} : (y, -1) \in N_{\text{epi}\,\varphi}(z)\}, \text{ where } z = (p, \varphi(p)).$$

In what follows, we assume that a solution to Problem (2.1) exists. We denote it by $(\hat{x}, \hat{u}, \hat{\mu})$. The hypotheses imposed on the data of the problem we consider in this chapter are stated as follows:

(H1) Function φ is Lipschitz continuous. Vector function f is Lipschitz continuous w.r.t. x for all u and for a.a. t, and $\mathscr{B}(U) \times \mathscr{L}(T)$-measurable, where $\mathscr{B}(U) \times \mathscr{L}(T)$ denotes the Cartesian product σ-algebra of $\mathscr{B}(U)$, the Borel σ-algebra over U, and $\mathscr{L}(T)$, the Lebesgue σ-algebra over T. The matrix-valued function G is Borel measurable and, moreover, continuous w.r.t. t for all u.

(H2) The mapping f satisfies the following two estimates.
(i) For any $c > 0$, there exists $\kappa_c \in \mathrm{L}_1(T; \mathbb{R})$ such that

$$|f(x, u, t) - f(y, u, t)| \leq \kappa_c(t)|x - y| \;\; \forall x, y \in cB_{\mathbb{R}^n},\; u \in U, \text{ a.a. } t \in T.$$

(ii) There exists a function $\alpha \in \mathrm{L}_1(T; \mathbb{R})$ such that

$$\sup_{u \in U} |f(\hat{x}(t), u, t)| \leq \alpha(t) \;\; \text{a.a. } t \in T.$$

2.3 Maximum Principle

In this section, the main result of this chapter is presented: the necessary optimality conditions in the form of the Pontryagin maximum principle for Problem (2.1).

Define the Hamilton–Pontryagin function

$$H(x, u, \psi, t) := \langle \psi, f(x, u, t) \rangle$$

and function $Q(u, \psi, t) := G^*(u, t)\psi$ which plays the same role as H but w.r.t. the impulsive part of the dynamics.

The maximum principle for Problem (2.1) takes the following form.

Theorem 2.1 *Assume that hypotheses (H1), (H2) hold, and let $(\hat{x}, \hat{u}, \hat{\mu})$ be an optimal control process in (2.1).*

Then, there exists a number $\lambda \geq 0$ and an absolutely continuous function $\psi : T \to \mathbb{R}^n$ such that $\lambda + \|\psi\|_C \neq 0$, and

$$\dot{\psi}(t) \in -\text{conv}\partial_x H(t) \ \ \ell\text{-}a.e., \tag{2.2}$$

$$(\psi(t_0), -\psi(t_1)) \in \lambda\partial\varphi(\hat{p}) + N_S(\hat{p}), \tag{2.3}$$

$$\max_{u \in U} H(u, t) = H(t) \ \ \ell\text{-}a.e., \tag{2.4}$$

$$\sup_{u \in U} \sigma_K(Q(u, t)) \leq 0 \ \ \forall t \in T, \tag{2.5}$$

$$\sigma_K(Q(t)) = 0 \ \ \hat{\mu}\text{-}a.e. \tag{2.6}$$

Above, as is adopted in this book: $\hat{p} = (\hat{x}(t_0), \hat{x}(t_1))$, and if some of the arguments of H, Q, or of any of their partial derivatives are omitted, then it means that the extremal values substitute the missing arguments everywhere, i.e., $H(t) = H(\hat{x}(t), \hat{u}(t), \psi(t), t)$, $H(u, t) = H(\hat{x}(t), u, \psi(t), t)$.

Let us briefly comment on these conditions. Equation (2.2) is the nonsmooth version of the adjoint equation (1.11) in which the convexification of the limiting subdifferential is required. The convex hull cannot be removed as well-known examples illustrate (see, e.g., Example 6.34 in [5]). Relation (2.3) is the nonsmooth transversality condition. The maximum condition splits into two parts. One part given by (2.4) includes the conventional control dynamics, while the other, given by (2.5) and (2.6), deals with the impulsive control component. When $K = [0, +\infty)$, Conditions (2.5) and (2.6) become

$$\sup_{u \in U} Q(u, t) \leq 0 \ \ \forall t \in T,$$

and

$$Q(t) = 0 \ \ \hat{\mu}\text{-a.e.}$$

respectively. When G does not depend on u, these conditions become (1.14) and (1.15), as obtained in the previous chapter.

The approach to the proof of this result consists of the following scheme. Necessary conditions are first derived for control processes associated with boundary points of the reachable set of the dynamic control system. The maximum principle follows directly from these in view of a simple reduction. Let us proceed with the details.

Firstly, let us make an observation. Note that it is not restrictive to consider the geometrically separated endpoint constraints, that is, $p \in S_0 \times S_1$, where S_0, S_1 are closed subsets of $\mathbb{R}^n$, rather than the case of joint constraints $p \in S$. Indeed, the general case is readily reduced to this particular one with separated constraints by introducing the new state variable $y \in \mathbb{R}^n$ such that $\dot{y} = 0$ and by imposing the new endpoint constraints:

$$(x_0, y_0) \in S, \quad x_1 = y_1.$$

Therefore, in what follows, we consider that $S = S_0 \times S_1$, as this convention simplifies the arguments of proof. Following the same rationale, we consider that φ depends solely on x_1 and not on $p = (x_0, x_1)$.

Let $\Psi : \mathbb{R}^n \to \mathbb{R}^d$ be a Lipschitz continuous function. We define the Ψ-attainable set, $\mathscr{A}_\Psi$, to be the set of all vectors $\Psi(x(t_1))$, where $x(\cdot)$ is some feasible trajectory, in the sense that (x, u, μ) is a control process for some control policy (u, μ) while $x(t_0) \in S_0$. A control process (x^*, u^*, μ^*) such that $x^*(t_0) \in S_0$ and $\Psi(x^*(t_1))$ is a boundary point of $\mathscr{A}_\Psi$, which is called a Ψ-boundary process. Note that the concepts of optimal process and of boundary process are, in fact, very close, as the arguments presented below will reveal.

The proof of the maximum principle is based on the following auxiliary assertion (necessary conditions for the boundary process).

Lemma 2.1 *Let the data satisfy hypotheses (H1), (H2), and let (x^*, u^*, μ^*) be a Ψ-boundary process.*

Then, there exist an absolutely continuous function ψ and a unit vector $e \in \mathbb{R}^d$ such that

$$\dot{\psi}(t) \in -\operatorname{conv} \partial_x H(x^*(t), u^*(t), \psi(t), t) \ \ell\text{-a.e.}, \tag{2.7}$$

$$\psi(t_0) \in N_{S_0}(x^*(t_0)), \quad \psi(t_1) \in -e \cdot \partial\Psi(x^*(t_1)), \tag{2.8}$$

$$\max_{u \in U} H(x^*(t), u, \psi(t), t) = H(x^*(t), u^*(t), \psi(t), t) \ \ell\text{-a.e.}, \tag{2.9}$$

$$\sup_{u \in U} \sigma_K(Q(u, \psi(t), t)) \le 0 \ \ \forall t \in T, \tag{2.10}$$

$$\sigma_K(Q(u^*(t), \psi(t), t)) = 0 \ \ \hat{\mu}\text{-a.e.} \tag{2.11}$$

The proof of this lemma is given in the next section. Now, let us demonstrate how Theorem 2.1 follows from the lemma.

Proof of Theorem 2.1. Let $(\hat{x}, \hat{u}, \hat{\mu})$ be an optimal process in Problem (2.1). Consider a new control system with constraints:

$$\begin{aligned}
&dx = f(x, u, t)dt + G(u, t)d\mu, \\
&\dot{y} = 0, \\
&\dot{z} = 0, \\
&x(t_0) \in S_0, \\
&y(t_0) \in S_1, \\
&z(t_0) \ge 0,
\end{aligned} \tag{2.12}$$

where y, z are auxiliary state variables with values in $\mathbb{R}^n$, $\mathbb{R}$. Define the mapping $\tilde{\Psi} : \mathbb{R}^{2n+1} \to \mathbb{R}^{n+1}$ as

$$\tilde{\Psi}(x, y, z) := \begin{bmatrix} z + \varphi(x) \\ y - x \end{bmatrix}$$

and let $R_{\tilde{\Psi}}$ denote the $\tilde{\Psi}$-attainable set for (2.12). Let us examine the control process $\pi := (\hat{x}, \hat{y}, \hat{z}, \hat{u}, \hat{\mu})$, where $\hat{y} \equiv \hat{x}(t_1)$ and $\hat{z} = 0$. In view of the optimality of process $(\hat{x}, \hat{u}, \hat{\mu})$ in Problem (2.1), the process π is $\tilde{\Psi}$-boundary to (2.12). Indeed, this follows from the fact that $L \cap R_{\tilde{\Psi}} = \varnothing$, where $L := \{(x, y, z) : x = y, z < \varphi(\hat{x}(t_1))\}$. Now, we apply Lemma 2.1 to process π. This is a straightforward task to verify that Conditions (2.2)–(2.6) are satisfied, and the assertion of Theorem 2.1 holds true.

The proof is complete. □

The given proof shows that the maximum principle follows from Lemma 2.1 by virtue of simple reduction. Thus, the main difficulty is the proof of this key-lemma.

2.4 Proof of Lemma 2.1

The proof of the key-lemma is preceded by a number of subsidiary results which will be required further on. These results concern an approximation technique which underlies the arguments of the proof and in which the measures driving the control system are replaced by conventional control functions.

Proposition 2.1 *Consider a sequence of maps $\zeta_i : \mathbb{R}^n \times T \to \mathbb{R}^n$, a sequence of measures $\nu_i \in C^*(T; \mathbb{R}^n)$, and a sequence of vectors $a_i \in \mathbb{R}^n$, $i = 1, 2, \ldots$, such that $\nu_i \overset{w*}{\to} \nu$, $a_i \to a$, and*

$$\ell\Big(\{t : \zeta_i(\cdot, t) = \zeta(\cdot, t)\}\Big) \to 1$$

as $i \to \infty$, for some map $\zeta : \mathbb{R}^n \times T \to \mathbb{R}^n$, measure $\nu \in C^*(T; \mathbb{R}^n)$, and vector $a \in \mathbb{R}^n$.

Let $\hat{x}(\cdot) : T \to \mathbb{R}^n$ be a function of bounded variation such that $\hat{x}(t_0) = a$, and

$$\hat{x}(t) = \hat{x}(t_0) + \int_{t_0}^{t} \zeta(\hat{x}(s), s)ds + \int_{[t_0,t]} d\nu \quad \forall t \in (t_0, t_1].$$

Assume that maps ζ_i satisfy the same hypotheses w.r.t. x, t imposed on f in (H1), and, uniformly w.r.t. i, the same estimates required in (H2).

Then, for each i sufficiently large, there exists a function $x_i(\cdot)$ of bounded variation such that $x_i(t_0) = a_i$,

$$x_i(t) = x_i(t_0) + \int_{t_0}^{t} \zeta_i(x_i(s), s)ds + \int_{[t_0,t]} d\nu_i \quad \forall t \in (t_0, t_1],$$

and $x_i \overset{w*}{\to} \hat{x}$ *while*[4]

$$x_i(t) - \int\limits_{[t_0,t]} d\nu_i \rightrightarrows \hat{x}(t) - \int\limits_{[t_0,t]} d\nu \ \text{ uniformly w.r.t. } t \in T. \tag{2.13}$$

Proof Define $\tilde{x}(t) := \hat{x}(t) - \int_{[t_0,t]} d\nu$. Then, $\tilde{x}(\cdot)$ is an absolutely continuous function which satisfies the differential equation

$$\tilde{x}'(t) := \frac{d\tilde{x}}{dt}(t) = \zeta\Big(\tilde{x}(t) + \int\limits_{[t_0,t]} d\nu, t\Big), \ \ \tilde{x}(t_0) = a. \tag{2.14}$$

Choose $\delta > 0$ such that $\|\tilde{x}\|_{\mathrm{C}} + \|\nu_i\| < \delta/2$ for all i sufficiently large. Let

$$\rho_i(t) := \left|\tilde{x}'(t) - \zeta_i\Big(\tilde{x}(t) + a_i - a + \int\limits_{[t_0,t]} d\nu_i, t\Big)\right|,$$

In view of (2.14),

$$\rho_i(t) = \left|\zeta\Big(\tilde{x}(t) + \int\limits_{[t_0,t]} d\nu, t\Big) - \zeta_i\Big(\tilde{x}(t) + a_i - a + \int\limits_{[t_0,t]} d\nu_i, t\Big)\right|.$$

Denote $T_i := \{t \in T : \zeta_i(x,t) = \zeta(x,t) \ \forall x \in \mathbb{R}^n\}$. For $t \in T \setminus T_i$, we have

$$\rho_i(t) \le \left|\zeta\Big(\tilde{x}(t) + \int\limits_{[t_0,t]} d\nu, t\Big)\right| + \left|\zeta_i\Big(\tilde{x}(t) + \int\limits_{[t_0,t]} d\nu, t\Big)\right| + \kappa_\delta(t)\gamma_i(t) \le$$

$$\le 2\alpha(t) + \kappa_\delta(t)\gamma_i(t), \ \text{ where } \gamma_i(t) := |a_i - a| + \int\limits_{[t_0,t]} d\nu_i - \int\limits_{[t_0,t]} d\nu,$$

and κ_δ, α are defined in (H2). For $t \in T_i$, again, from (H2),

$$\rho_i(t) \le \kappa_\delta(t)\gamma_i(t).$$

In view of the imposed assumptions, γ_i are uniformly bounded in L_∞-norm and $\gamma_i \to 0$ a.e. Therefore, the application of the dominated convergence theorem yields that $|\rho_i|_{\mathrm{L}_1} \to 0$ as $i \to \infty$. Then, from the proof of Theorem 3.1.6 in [1], it follows that there exists an absolutely continuous function $\tilde{x}_i$ satisfying

[4]Weak-* convergence in $\mathrm{BV}(T; \mathbb{R}^n)$.

$$\tilde{x}_i'(t) = \zeta_i\Big(\tilde{x}_i(t) + \int\limits_{[t_0,t]} d\nu_i, t\Big), \quad \tilde{x}_i(t_0) = a_i,$$

for i sufficiently large. Moreover,

$$\tilde{x}_i(t) - \tilde{x}(t) \rightrightarrows 0 \text{ uniformly w.r.t. } t \in T, \tag{2.15}$$

and

$$\int\limits_{t_0}^{t_1} |\tilde{x}_i'(t) - \tilde{x}'(t)|dt \to 0. \tag{2.16}$$

Now define $x_i(t) := \tilde{x}_i(t) + \int_{[t_0,t]} d\nu_i$ for $t > t_0$, and $x_i(t_0) = \tilde{x}_i(t_0)$. Property (2.15) gives precisely (2.13). Property (2.16), together with the fact that $\nu_i \overset{w*}{\to} \nu$, implies that $x_i \overset{w*}{\to} \hat{x}$. The proof is complete. □

Proposition 2.2 *Take $\nu \in C^*(T; \mathbb{R}^k)$ and $\{\nu_i\}$ with $\nu_i \in C^*(T; \mathbb{R}^k)$. Let $\|\nu_i\| \leq$ const, and $F(t; \nu_i) \to F(t; \nu)$ for a.a. $t \in T$ including $t = t_1$, as $i \to \infty$. Then, $\nu_i \overset{w*}{\to} \nu$.*

Conversely, let $\nu_i \overset{w*}{\to} \nu$. Then, there exists a complement of a countable set D with $t_1 \in D$ and also a subsequence $\{\nu_{i_j}\}$ of $\{\nu_i\}$ such that

$$F(t; \nu_{i_j}) \to F(t; \nu) \;\; \forall t \in D.$$

The proof of this proposition is left for the Reader as a simple exercise, Exercise 2.3.

The next proposition is somewhat technical and requires more effort.

Proposition 2.3 *Let $g : \mathbb{R}^m \times T \to \mathbb{R}^n$ be a Borel measurable function which satisfies the same hypotheses imposed on G in (H1). Take $\mu \in C^*(T; \mathbb{R})$, $\mu \geq 0$, and a Borel measurable function $u(\cdot)$ such that $u(t) \in U$ for a.a. t w.r.t. $\ell + \mu$, and $r(u(t), t)$ is μ-integrable.*

Then, there exist a sequence of Borel measurable functions $u_i : T \to \mathbb{R}^m$, and a sequence of functions $m_i \in L_1(T; \mathbb{R})$ such that, for all i, $u_i(t) \in U$, $m_i(t) \geq 0$ for a.a. t w.r.t. ℓ, the function $r(u_i(t), t)m_i(t)$ is integrable w.r.t. ℓ, and $\mu_i \overset{w*}{\to} \mu$, $\eta_i \overset{w*}{\to} \eta$, where

$$d\mu_i = m_i(t)dt, \quad d\eta_i = r(u_i(t), t)d\mu_i, \quad d\eta = r(u(t), t)d\mu,$$

and

$$\ell\Big(\{t : u(t) \neq u_i(t)\}\Big) \to 0 \;\; \text{as } i \to \infty.$$

Proof The analysis of the statement of this proposition suggests that it is not restrictive to consider that μ is merely singular w.r.t. ℓ. Then, there exists a sequence of open subsets $A_i \subset T$ such that supp $\mu \subset A_i$ $\forall i = 1, 2, \ldots$, and $\ell(A_i) \to 0$ as $i \to \infty$.

For $j = 0, 1, \ldots, i-1$ and $i = 1, 2, \ldots$, define the intervals I_{ij} and A_{ij}:

$$I_{ij} := \left[\frac{j}{i}, \frac{j+1}{i}\right), \quad A_{ij} := I_{ij} \cap A_i.$$

Define $\hat{m}_{ij} \in \mathbb{R}_+$ by the formula

$$\hat{m}_{ij} := \begin{cases} \dfrac{\mu(A_{ij})}{\ell(A_{ij})}, & \text{when } A_{ij} \neq \varnothing, \\ 0, & \text{otherwise} \end{cases}$$

and $m_{i,j} : I_{ij} \to \mathbb{R}_+$ by the formula

$$m_{ij}(t) := \begin{cases} \hat{m}_{ij}, & \text{when } t \in A_{ij}, \\ 0, & \text{otherwise.} \end{cases}$$

We also define $\hat{\gamma}_{ij} \in \mathbb{R}^n$ as

$$\hat{\gamma}_{ij} := \frac{\eta(A_{ij})}{\mu(A_{ij})} \quad \text{if } \hat{m}_{ij} \neq 0$$

and $\tilde{\gamma}_{ij} : I_{ij} \to \mathbb{R}^n$ as

$$\tilde{\gamma}_{ij}(t) := \begin{cases} \hat{\gamma}_{ij}, & \text{when } t \in A_{ij} \text{ and } \hat{m}_{ij} \neq 0, \\ r(u(t), t), & \text{otherwise.} \end{cases}$$

Next, define $\gamma_{ij} : I_{ij} \to \mathbb{R}^n$ by

$$\gamma_{ij}(t) := \begin{cases} \xi_{ij}(t), & \text{when } t \in A_{ij} \text{ and } \hat{m}_{ij} \neq 0, \\ r(u(t), t), & \text{otherwise,} \end{cases}$$

where the function ξ_{ij} is uniquely specified by

$$\xi_{ij}(t) \in \operatorname*{argmin}_{\xi \in \mathrm{cl\,conv}\, r(U,t)} |\xi - \hat{\gamma}_{ij}|.$$

(Note that the above right-hand side is a single-point set.)

Observe that ξ_{ij} is continuous on its domain. This readily follows from the continuity property imposed on the map r w.r.t. t-argument. Consequently, γ_{ij} is a Borel measurable function. Moreover, from its definition, and appealing again to the continuity property, we deduce existence of a sequence $\{\varepsilon_i\}$ of real numbers converging to zero such that

$$|\tilde{\gamma}_{ij}(t) - \gamma_{ij}(t)| \leq \varepsilon_i \quad \forall i, j, \quad \text{and } t \in I_{i,j}. \tag{2.17}$$

Now, define two Borel functions $m_i : T \to \mathbb{R}$, and $\gamma_i : T \to \mathbb{R}^n$ by

$$m_i(t) := \sum_j m_{ij}(t)\chi_{I_{ij}}(t), \quad \gamma_i(t) := \sum_j \gamma_{ij}(t)\chi_{I_{ij}}(t),$$

where χ_A denotes the characteristic, or indicator, function of set A, that is:

$$\chi_A(t) := \begin{cases} 1, & \text{if } t \in A, \\ 0, & \text{if } t \notin A. \end{cases}$$

Let us ensure that the m_i's have the properties asserted in the proposition. Clearly, function $r(u_i(t), t)m_i(t)$ is ℓ-integrable on T. It is immediately apparent that the m_i's are nonnegative and bounded. Moreover, it is a straightforward task to verify that constructed measures μ_i: $d\mu_i = m_i(t)dt$ converge weakly-* to μ (see Exercise 2.4).

Our next objective is to show that

$$\tilde{\eta}_i \overset{w*}{\to} \eta, \quad \text{where } d\tilde{\eta}_i = \gamma_i(t)m_i(t)dt. \tag{2.18}$$

Take arbitrary $\phi \in C(T; \mathbb{R}^n)$. We have:

$$\int_{t_0}^{t_1} \phi(t)\gamma_i(t)m_i(t)dt = \sum_{j\in J_i} \int_{A_{ij}} \phi(t)\tilde{\gamma}_{ij}(t)m_{ij}(t)dt$$
$$+ \sum_{j\in J_i} \int_{A_{ij}} \phi(t)[\gamma_{ij}(t) - \tilde{\gamma}_{ij}(t)]m_{ij}(t)dt,$$

where J_i stands for the set of indices j such that $\hat{m}_{ij} \neq 0$. The second term on the right-hand side is bounded by $\varepsilon_i \|\phi\|_C \|\mu\|$ in view of (2.17). Hence, it tends to zero as $i \to \infty$. Consider the first term, that is

$$\sum_{j\in J_i} \int_{A_{ij}} \phi(t)\tilde{\gamma}_{ij}(t)m_{ij}(t)dt.$$

According to the definitions, we express it as

$$\sum_{j\in J_i} \int_{A_{ij}} \phi(t) \cdot \frac{\eta(A_{ij})}{\mu(A_{ij})} \cdot \frac{\mu(A_{ij})}{\ell(A_{ij})} \cdot dt = \sum_{j\in J_i} \int_{A_{ij}} \phi(t)(\ell(A_{ij}))^{-1}dt \cdot \int_{A_{ij}} r(u(s), s)d\mu$$
$$= \sum_{j\in J_i} \int_{A_{ij}} \phi(s)r(u(s), s)d\mu + \sum_{j\in J_i} \int_{A_{ij}} \Big(\int_{A_{ij}} \phi(t)(\ell(A_{ij}))^{-1}dt - \phi(s)\Big) r(u(s), s)d\mu.$$

The second term in the last expression converges to zero as $i \to \infty$ due to the mean value theorem and also to the fact that $\phi(\cdot)$ is uniformly continuous (see Exercise 2.5). Then, (2.18) is proved, as $\operatorname{supp}\mu \in A_i$.

Define the functions $h_i : \mathbb{R}^m \times T \to \mathbb{R}^n$ and the set-valued mappings U_i on T as

$$h_i(v, t) := r(v, t)m_i(t), \quad U_i(t) := \begin{cases} U, & \text{if } t \in A_i, \\ \{u(t)\}, & \text{otherwise.} \end{cases}$$

Note that by construction $\gamma_i(t)m_i(t) \in \operatorname{cl}\operatorname{conv} h_i(U_i(t), t)$ for all $t \in T$.

Let $\{\delta_i\}$ be an arbitrary sequence of positive numbers such that $\delta_i \to 0$. A routine argument involving the application of Aumann's theorem (see, e.g., [1], Theorem 3.1.3) to the set-valued map $t \to \operatorname{cl} h_i(U_i(t), t)$ allows one to conclude that there exists a measurable function $d_i : T \to \mathbb{R}^n$ such that $d_i(t) \in \operatorname{cl} h_i(U_i(t), t)$ for a.a. $t \in T$, and

$$\sup_{t\in T}\left|\int_{t_0}^{t} [d_i(s) - \gamma_i(s)m_i(s)]ds\right| < \frac{\delta_i}{2}, i = 1, 2, \ldots$$

In view of the measurable selection lemma (Filippov's lemma, [3]), there also exists a measurable function $u_i(\cdot)$ such that $u_i(t) \in U_i(t)$ for a.a. $t \in T$, and

$$\sup_{t\in T}\left|\int_{t_0}^{t} [d_i(s) - h_i(u_i(s), s)]ds\right| < \frac{\delta_i}{2}, \quad i = 1, 2, \ldots$$

Then,

$$\sup_{t\in T}\left|\int_{t_0}^{t} \gamma_i(s)m_i(s)ds - \int_{t_0}^{t} r(u_i(s), s)m_i(s)]ds\right| < \delta_i,$$

and $u_i(t) = u(t)$ for a.a.$t \in T \setminus A_i$, $u_i(t) \in U$ for a.a.$t \in A_i$. This estimate, in view of (2.18) and Proposition 2.2, implies that $\eta_i \overset{w*}{\to} \eta$. Since $\ell(A_i) \to 0$, the constructed functions $u_i(\cdot)$ have the properties asserted in the lemma. The proof is complete. □

Proof of Lemma 2.1. In the first stage, we assume that the control measure μ is scalar-valued, $K = [0, \infty)$, and, moreover, that the matrix-valued function G is bounded on $U \times T$. Later, it will be shown that such assumptions are not restrictive. Since μ is scalar, henceforth, g is written in place of G to stress the fact that the map G is now vector-valued, not matrix-valued.

The fact that $\Psi(x^*(t_1)) \in \partial\mathscr{A}_\Psi$ implies that there exists a sequence $\{\xi_i\}$ such that $\xi_i \notin \mathscr{A}_\Psi$ and $\xi_i \to \Psi(x^*(t_1))$. By means of Propositions 2.1 and 2.3, there exists a sequence of control processes $(\bar{x}_i, \bar{u}_i, \bar{m}_i)$, where Lebesgue measurable functions $\bar{u}_i$ and $\bar{m}_i$ are such that $\bar{u}_i(t) \in U$ ℓ-a.e. and $\bar{m}_i \in L_1(T; \mathbb{R})$, satisfying

$$\begin{cases} \bar{x}_i'(t) = f(\bar{x}_i(t), \bar{u}_i(t), t) + g(\bar{u}_i(t), t)\bar{m}_i(t), \quad \text{a.a. } t, \\ \bar{x}_i(t_0) = x^*(t_0), \\ \bar{\mu}_i \overset{w*}{\to} \mu^*, \\ \bar{x}_i \overset{w*}{\to} x^*, \\ \ell(\{t \in T : \bar{u}_i(t) \neq u^*(t)\}) \to 0, \end{cases} \tag{2.19}$$

where $d\bar{\mu}_i = \bar{m}_i(t)dt$.

From Proposition 2.2, by restricting attention to an appropriate subsequence, it can be arranged that $\bar{x}_i(t) \to x^*(t)$ on a full measure subset of T containing points t_0 and t_1. In view of Exercise 1.3, the uniform boundedness principle and the weak-* convergence, the functions $|\bar{x}_i(\cdot)|$ are majorized by a common constant function. Then, from the Lebesgue dominated convergence theorem, it follows that $\|\bar{x}_i - x^*\|_{L_2} \to 0$ as $i \to \infty$.

Define, for $i = 1, 2, \ldots$, the number ε_i as

$$\varepsilon_i := \left(\int_{t_0}^{t_1} |\bar{x}_i(t) - x^*(t)|^2 dt + |\xi_i - \Psi(\bar{x}_i(t_1))| + |\bar{x}_i(t_1) - x^*(t_1)|^2 \right)^{\frac{1}{2}},$$

and the function

$$K_i(t) := i + \max_{j \le i}\{\bar{m}_j(t)\}.$$

It is clear that $\varepsilon_i \to 0$ and that each K_i majorizes $\bar{m}_i$, while the sequence $\{K_i\}$ is monotone with pointwise limit $+\infty$.

For each i, consider the conventional that is nonimpulsive, optimal control problem

$$\begin{aligned}
&\text{Minimize } |\xi_i - \Psi(x(t_1))| + |x(t_1) - x^*(t_1)|^2 + \int_{t_0}^{t_1} |x(t) - x^*(t)|^2 dt \\
&\text{subject to } \dot{x}(t) = f(x(t), u(t), t) + g(u(t), t)m(t), \\
&\qquad x(t_0) \in S_0, \\
&\qquad (u(t), m(t)) \in U \times [0, K_i(t)] \ \ell\text{-a.e.},
\end{aligned} \tag{2.20}$$

in which m is regarded as a component in the joint control variable (u, m).

Problem (2.20) can be reformulated as follows

$$\text{Minimize } \ \Phi_i(a, u, m) \ \text{ over } \ (a, u, m) \in M_i,$$

where

$$\Phi_i(a, u, m) := |\xi_i - \Psi(x(t_1))| + |x(t_1) - x^*(t_1)|^2 + \int_{t_0}^{t_1} |x(t) - x^*(t)|^2 dt,$$

$$M_i := \{(a, u, m) : \ a \in S_0, \ u(t) \in U, \ m(t) \in [0, K_i(t)] \text{ a.a. } t \in T\},$$

functions $u(\cdot)$, $m(\cdot)$ are ℓ-measurable, $x(\cdot)$ is the trajectory associated with (a, u, m).

Let us endow M_i with the metric ρ defined by

$$\rho((a,u,m),(\tilde{a},\tilde{u},\tilde{m})) := |a-\tilde{a}| + \ell\Big(\{t\in T : u(t)\neq \tilde{u}(t)\}\Big) + \int_{t_0}^{t_1} |m(t)-\tilde{m}(t)|dt.$$

This is a straightforward task to ensure that the pair (M_i, ρ) represents a complete metric space; see Exercise 2.6. From Proposition 2.1, it follows that the mapping Φ_i is continuous for this choice of metric.

Observe that, for each i,

$$\Phi_i(x^*(t_0), \bar{u}_i, \bar{m}_i) \le \inf_{(a,u,m)\in M_i} \Phi_i(a,u,m) + \varepsilon_i^2.$$

Then, since $\Phi_i \ge 0$, it follows from Ekeland's variational principle, [2], that, for each i, there exists a control process $(a_i, u_i, m_i) \in M_i$ such that

$$\Phi_i(a_i,u_i,m_i) \le \Phi_i(a,u,m) + \varepsilon_i \rho((a_i,u_i,m_i),(a,u,m)) \;\; \forall\,(a,u,m)\in M_i, \tag{2.21}$$

$$\rho((a_i,u_i,m_i),(x^*(t_0),\bar{u}_i,\bar{m}_i)) \le \varepsilon_i. \tag{2.22}$$

Let x_i be the trajectory associated with control process (a_i, m_i, u_i). From (2.19), (2.22), it follows that

$$\begin{cases} x_i(t_0) \to x^*(t_0), \\ \ell(\{t : u_i(t) \neq u^*(t)\}) \to 0, \\ \mu_i \overset{w*}{\to} \mu^*, \end{cases} \tag{2.23}$$

where $d\mu_i = m_i(t)dt$.

From (2.21), (2.22), we conclude that $\Phi_i(a_i,u_i,m_i) \to 0$ as $i\to\infty$. This implies that $\|x_i - x^*\|_{L_2} \to 0$ and that $|x_i(t_1) - x^*(t_1)| \to 0$. Then, there exists a subsequence for which

$$x_i(t) \to x^*(t) \;\; \forall\, t \in D \cup \{t_0\} \cup \{t_1\}, \tag{2.24}$$

where D is some full Lebesgue measure subset of T.

It is not difficult to show (by application of Gronwall's inequality, Exercise 1.2) that $\mathrm{Var}\,|_T x_i \le \mathrm{const}\;\forall\, i$. Hence, by Proposition 2.2, $x_i \overset{w*}{\to} x^*$. In view of (2.23), (2.24), and by using the dominated convergence theorem, we obtain

$$\eta_i \overset{w*}{\to} \eta, \;\; \text{where} \;\; d\eta_i = g(u_i(t),t)m_i(t)dt, \;\; d\eta = g(u^*(t),t)d\mu^*. \tag{2.25}$$

Inequality (2.21) means that the process (x_i, u_i, m_i) is optimal in the following control problem:

$$\text{Minimize } |\xi_i - \Psi(x(t_1))| + |x(t_1) - x^*(t_1)|^2 + \int_{t_0}^{t_1} |x(t) - x^*(t)|^2 dt +$$
$$\varepsilon_i \left(|x(t_0) - x_i(t_0)| + \int_{t_0}^{t_1} |m(t) - m_i(t)| dt + \int_{t_0}^{t_1} \chi_i(u(t), t) dt \right) \tag{2.26}$$
$$\text{subject to } \dot{x} = f(x, u, t) + g(u, t)m,$$
$$x(t_0) \in S_0,$$
$$(u(t), m(t)) \in U \times [0, K_i(t)] \ \text{ a.a. } t \in T.$$

where $\chi_i(u,t) = 0$ if $u = u_i(t)$, and $\chi_i(u,t) = 1$ otherwise.

For each i, Problem (2.26) is a conventional optimal control problem, for which necessary conditions of optimality are available. The application of nonsmooth version of the maximum principle, [5, 7], yields the existence of an absolutely continuous function $\psi_i : T \to \mathbb{R}^n$ such that[5]

$$\dot{\psi}_i(t) \in -\text{conv}\partial_x H(x_i(t), u_i(t), \psi_i(t), t) + 2(x_i(t) - x^*(t)), \tag{2.27}$$
$$\max_{(u,m)\in U\times[0,K_i(t)]} H_i(u, m, t) = H_i(u_i(t), m_i(t), t), \tag{2.28}$$
$$\psi_i(t_0) \in N_{S_0}(x_i(t_0)) + \varepsilon_i B_{\mathbb{R}^n}, \tag{2.29}$$
$$\psi_i(t_1) \in -\partial_x |\xi_i - \Psi(x)| \Big|_{x=x_i(t_1)} - 2(x_i(t_1) - x^*(t_1)), \tag{2.30}$$

where, for convenience, it is defined

$$H_i(u, m, t) := \Big\langle \psi_i(t), f(x_i(t), u, t) + g(u, t)m \Big\rangle - \varepsilon_i \Big(\chi_i(u, t) + |m - m_i(t)| \Big).$$

In view of the fact that $\xi_i \neq \Psi(x_i(t_1))$, it follows from [4] that (2.30) implies

$$\psi_i(t_1) \in -e_i \cdot \partial\Psi(x_i(t_1)) - 2(x_i(t_1) - x^*(t_1)) \tag{2.31}$$

for some unit vector $e_i \in \mathbb{R}^n$.

By using Gronwall's inequality, it is deduced from (2.27) and (2.31) that the ψ_i's are a uniformly bounded equicontinuous family of functions. Extraction of a subsequence ensures that $\psi_i(t) \rightrightarrows \psi(t)$ uniformly in T, where ψ is some continuous function.

Let us show that ψ has all the properties asserted in the lemma.

The routine arguments encompass the derivation of the adjoint differential inclusion from (2.27), that is, (2.7), and also the fact that function ψ is absolutely continuous. By passing to a subsequence, the existence of a unit vector e is obtained such that $e_i \to e$. The transversality conditions (2.8) follow from (2.29) and (2.31) in view of the upper semicontinuity of the limiting normal cone and subdifferential, [4].

Next, we show that the conditions (2.9) and (2.10) are satisfied.

---

[5] Note that (2.26) is a free right endpoint problem, and therefore, its extremals are normal.

The following simple observation will be used repeatedly: If $\{\Sigma_i\}$ is a sequence of ℓ-measurable subsets of T and $\ell(\Sigma_i) \to 0$, then it is possible to replace the sequence $\{\Sigma_i\}$ with a subsequence (without relabeling) with the following property:

$$\ell\Big(\{t \in T \setminus \Sigma_j \ \forall j \geq i\}\Big) \to 1 \ \text{ as } i \to \infty.$$

Define

$$\begin{aligned} A_i &:= \Big\{t \in T : \big\langle \psi_i(t), g(u_i(t), t)\big\rangle > \varepsilon_i\Big\}, \\ B_i &:= \Big\{t \in T : m_i(t) > \frac{1}{\sqrt{\varepsilon_i}}\Big\}, \\ C_i &:= \Big\{t \in T : \sup_{u \in U} |H(x_i(t), u, \psi_i(t), t)| > \sqrt{K_i(t)}\Big\}. \end{aligned}$$

Note that

$$\ell\Big(A_i \cup B_i \cup C_i\Big) \to 0 \ \text{ as } i \to \infty.$$

Indeed, $\ell(A_i) \to 0$ as the maximization in (2.28) implies that $m_i(t) = K_i(t)$ for a.a. $t \in A_i$, while the K_i's increase uniformly to infinity. The fact that the sequence $\{m_i\}$ is norm bounded in L_1 implies that $\ell(B_i) \to 0$. Finally, $\ell(C_i) \to 0$ in view of the uniform integrable boundedness condition as stated in (H2).

Let $T_i \subset T$ be the set of points t such that

$$\big\langle \psi_j(t), g(u_j(t), t)\big\rangle \leq \varepsilon_j, \tag{2.32}$$

$$\sup_{u \in U} |H(x_j(t), u, \psi_j(t), t)| \leq \sqrt{K_j(t)}, \tag{2.33}$$

$$m_j(t) \leq \frac{1}{\sqrt{\varepsilon_j}}, \tag{2.34}$$

$$u_j(t) = u^*(t), \tag{2.35}$$

$$H_j(u_j(t), m_j(t), t) = \max_{(u,m) \in U \times [0, K_j(t)]} H_j(u, m, t), \tag{2.36}$$

for all $j \geq i$, and

$$\lim_{j \to \infty} x_j(t) \to x^*(t) \ \text{ as } j \to \infty. \tag{2.37}$$

In view of the earlier observations, it holds, by extraction of a subsequence, that $\ell(T_i) \to 1$. Take any $t \in \cup_i T_i$ and any $\bar{u} \in U$. For some i, it holds that $t \in T_i$, whence, by (2.35) and (2.36), $H_j(u^*(t), m_j(t), t) \geq H_j(\bar{u}, 0, t)$, $\forall j \geq i$. This inequality together with (2.32) and (2.34) gives

$$H(x_j(t), u^*(t), \psi_j(t), t) + \frac{\varepsilon_j}{\sqrt{\varepsilon_j}} \geq H(x_j(t), \bar{u}, \psi_j(t), t) - \varepsilon_j - \frac{\varepsilon_j}{\sqrt{\varepsilon_j}}, \quad \forall j \geq i.$$

In view of (2.37), as $j \to \infty$, we obtain

$$H(x^*(t), u^*(t), \psi(t), t) \geq H(x^*(t), \bar{u}, \psi(t), t).$$

Thus, (2.9) is established on the subset $\cup_i T_i$ of full ℓ-measure.

Again, take any $t \in \cup_i T_i$. For each j, let $\bar{u}_j$ be chosen such that

$$Q(\bar{u}_j, \psi_j(t), t) > \sup_{u \in U} Q(u, \psi_j(t), t) - \varepsilon_i. \tag{2.38}$$

From (2.35), (2.36), it follows, for some i, that $H_j(u^*(t), m_j(t), t) \geq H_j(\bar{u}_j(t), K_j(t), t)$ for all $j \geq i$. This, together with (2.32)–(2.34), and (2.38) gives

$$\sqrt{K_j(t)} + \sqrt{\varepsilon_j} \geq -\sqrt{K_j(t)} + \sup_{u \in U} Q(u, \psi_j(t), t) K_j(t) - \varepsilon_j - 2\varepsilon_j K_j(t), \quad \forall\, j \geq i.$$

However, $K_j(t) \to \infty$ as $j \to \infty$. Dividing across this inequality by $K_j(t)$ and passing to the limit as $j \to \infty$, we obtain the inequality

$$\sup_{u \in U} Q(u, \psi(t), t) \leq 0,$$

which holds on a full measure subset of T. Then, in view of continuity of g w.r.t. t, Condition (2.10) is proved.

It remains to ensure that (2.11) is satisfied. By putting $u = u_i(t)$ in the maximum condition (2.28), we obtain

$$Q(u_i(t), \psi_i(t), t) m_i(t) \geq Q(u_i(t), \psi_i(t), t) m - \varepsilon_i |m - m_i(t)| \;\; \forall m \in [0, K_i(t)], \quad \text{a.a. } t.$$

When $m_i(t) > 0$, this inequality implies that

$$Q(u_i(t), \psi_i(t), t) > -\varepsilon_i. \tag{2.39}$$

Therefore, (2.39) is valid ℓ-a.e. on the set $L_i := \{t \in T : m_i(t) > 0\}$.

Define ν_i and ν from $C(T; \mathbb{R})$ by

$$d\nu_i = Q(u_i(t), \psi_i(t), t) d\mu_i, \quad d\nu = Q(u^*(t), \psi(t), t) d\mu^*.$$

By means of (2.25) we have $\nu_i \overset{w*}{\to} \nu$. According to Proposition 2.2, we can arrange, by a subsequence extraction, that $F(t; \nu_i) \to F(t; \nu)$ for all t in some dense subset D_0 of T which contains $t = t_1$. For any $t \in D_0$ we have, by (2.39),

$$F(t;\nu_i)=\int\limits_{[t_0,t]} d\nu_i=\int\limits_{t_0}^{t} Q(u_i(s),\psi_i(s),s)m_i(s)ds=$$

$$=\int\limits_{[t_0,t]\cap L_i} Q(u_i(s),\psi_i(s),s)m_i(s)ds\geq -\varepsilon_i\int\limits_{[t_0,t]\cap L_i} m_i(s)ds=$$

$$=-\varepsilon_i\int\limits_{[t_0,t]} d\mu_i\geq -\varepsilon_i\|\mu_i\|.$$

The sequence $\{\|\mu_i\|\}$ is bounded, so $F(t;\nu)\geq 0$. Since the sets $[t_0,t]\cap D_0$ generate the Borel σ-algebra over T, we have

$$\int\limits_B Q(u^*(t),\psi(t),t)d\mu^*\geq 0\ \ \forall B\in\mathscr{B}(T).$$

Then, $Q(u^*(t),\psi(t),t)\geq 0$ μ^*-a.e. This inequality coupled with (2.10) yields (2.11).

In order to conclude the proof, it remains to justify that the simplifying hypotheses regarding K and G made in the beginning of the proof can be dispensed with. Let K now be an arbitrary convex closed cone in $\mathbb{R}^k$ while G is allowed to take unbounded values. Let us set $\tilde{K}:=\{w\in K:\sum_{i=1}^k|w^i|\leq 1\}$ and

$$\tilde{g}(u,w,t):=\frac{G(u,t)w}{1+\sum\limits_{i=1}^{n}|G_i(u,t)w|}$$

where G_i stands for the i-th row of G.

Consider a new dynamic system (reduction to a new problem):

$$\begin{cases} dx=f(x,u,t)dt+\tilde{g}(u,w,t)d\nu,\\ (u(t),w(t))\in U\times\tilde{K}\quad(\ell+\nu)\text{-a.e.},\\ x(t_0)\in S_0,\\ \nu\geq 0,\end{cases}\tag{2.40}$$

in which the control measure ν is scalar and the Borel control u is extended to (u,w) with $w\in\mathbb{R}^k$. Note that the original dynamic control system defined in (2.1) and the new one defined by (2.40) are equivalent, since they generate the same set of trajectories; see Exercise 2.7.

We define w^* to be the Radon–Nikodym derivative of μ^* w.r.t. $|\mu^*|$. Then, $w^*(t)\in\tilde{K}$ $|\mu^*|$-a.e. Define $\nu^*\in C^*(T;\mathbb{R})$ to be

$$d\nu^*=\Big(1+\sum_{i=1}^{n}|G_i(u^*(t),t)w^*(t)|\Big)d|\mu^*|.$$

Then, in view of the fact that Problems (2.1) and (2.40) are equivalent, the control process (x^*, u^*, w^*, ν^*) is also a Ψ-boundary control process for (2.40). Application of the special case of this lemma, already proved above, to (2.40) yields the full version of the theorem, which is a straightforward task to verify.

The proof is complete. □

2.5 Exercises

Exercise 2.1 Recall and prove the following basic facts from the measure theory.

(a) Any ℓ-measurable set $A \subseteq \mathbb{R}$ can be represented as

$$A = \left(\bigcup_{i=1}^{\infty} C_i\right) \cup Z,$$

where C_i are closed sets and $\ell(Z) = 0$. This, in particular, means that any ℓ-measurable set is the union of a Borel set and a zero measure set.

(b) Any ℓ-measurable function $f : T \to \mathbb{R}$ is ℓ-a.e. identical to a Borel function.

Exercise 2.2 The solution concept given by (2.1) is not well posed w.r.t. weak-* convergence of the measures. Construct an appropriate counterexample.

Exercise 2.3 Prove Proposition 2.2.

Exercise 2.4 Show that, for the measures μ_i constructed in the proof of Proposition 2.3, it holds that $\mu_i \overset{w*}{\to} \mu$.

Exercise 2.5 Function $f : A \to \mathbb{R}^k$, where $A \subseteq \mathbb{R}^n$, is said to be uniformly continuous on A, provided that $\forall \varepsilon > 0\ \exists \delta > 0$: $|f(x_1) - f(x_2)| \le \varepsilon\ \forall x_1, x_2 \in A$: $|x_1 - x_2| \le \delta$. Prove that any continuous function on a compact set is uniformly continuous.

Exercise 2.6 Regarding the metric ρ introduced in the proof of Lemma 2.1, show that

(a) function ρ is indeed a metric;

(b) set M_i endowed with ρ is a complete metric space.

Exercise 2.7 Show that the dynamic control systems defined in (2.1) and in (2.40) are equivalent. That is, the sets of the feasible trajectories coincide.

References

1. Clarke, F.: Optimization and Nonsmooth Analysis. Wiley-Interscience, New York (1983)
2. Ekeland, I.: On the variational principle. J. Math. Anal. Appl. **47**, 324–353 (1974)
3. Filippov, A.: On certain problems of optimal regulation. Bull. Moscow State Univ. Ser. Math. Mech. (2), 25–38 (1959)
4. Mordukhovich, B.: Variational analysis and generalized differentiation I. Basic theory. In: Grundlehren der Mathematischen Wissenschaften [Fundamental Principles of Mathematical Sciences], vol. 330. Springer, Berlin (2006)
5. Mordukhovich, B.: Variational analysis and generalized differentiation II. Applications. In: Grundlehren der Mathematischen Wissenschaften [Fundamental Principles of Mathematical Sciences], vol. 331. Springer, Berlin (2006)
6. Rockafellar, R., Wets, R.B.: Variational Analysis. Grundlehren der Math. Wissenschaften. Springer (1998)
7. Vinter, R.: Optimal Control. Birkhäuser, Boston (2000)
8. Vinter, R., Pereira, F.: Maximum principle for optimal processes with discontinuous trajectories. SIAM J. Control Optim. **26**(1), 205–229 (1988)

Chapter 3
Impulsive Control Problems Under the Frobenius Condition

Abstract In this chapter, the matrix-multiplier G for the impulsive control is enriched with a dependence on the state variable x. This naturally leads to some ambiguity in the choice of the state trajectory, since it is assumed that the state trajectory may exhibit jumps. Therefore, generally speaking, different types of integral w.r.t. measure will lead us to different solution concepts. Herein, we have settled on the type of integration which implies the stability of the solution w.r.t. approximations by absolutely continuous measures. The uniqueness and stability of the solution in this case are guaranteed by the well-known Frobenius condition. The extension of the original problem is treated w.r.t. this type of solution which is stable in the weak-* topology. The main result of this chapter is the second-order necessary conditions of optimality without a priori assumptions of normality, which are obtained under the assumption that the Frobenius condition for the columns of matrix G is satisfied. The chapter ends with 11 exercises.

3.1 Introduction

In the two previous chapters, the matrices $G(t)$ and $G(u, t)$ multiplying the Borel control measure have been considered. In this chapter, the case in which the matrix-multiplier G depends on the state variable and time, that is, $G = G(x, t)$, is considered. As will follow from the forthcoming considerations of this and next chapters, starting from this point, the extension procedure begins to encounter certain challenges and, thereby, it is not as simple as in Chaps. 1 and 2. Indeed, the key point is to ensure robustness of the impulsive control system w.r.t. the control measure by examining its approximations in the weak-* topology. Note that such approximations of measures by conventional controls naturally arise in applications. However, the robustness property is generally lost unless some extra assumptions on matrix G w.r.t. the x-variable are imposed. It appears that the most general assumption under which the robustness is still valid is the so-called Frobenius condition, presented and discussed in the next sections. Thus, this condition is assumed to be a priori

© Springer Nature Switzerland AG 2019
A. Arutyunov et al., *Optimal Impulsive Control*, Lecture Notes in Control and Information Sciences 477, https://doi.org/10.1007/978-3-030-02260-0_3

satisfied while the approach of this chapter is based on the application of the Frobenius theorem on global solvability of a system of partial differential equations.

In this chapter, the maximum principle and the second-order necessary conditions of optimality for an impulsive control problem are investigated under the Frobenius condition. One of the main features of these conditions is that no a priori normality assumptions are required. This feature follows from the fact that these conditions rely on an extremal principle, which is proved for an abstract minimization problem with equality and inequality constraints and constraints given by an inclusion in a convex cone. The proof of these conditions is based on a nonlinear transformation of the initial problem into another problem, in which G does not depend on x and for which the first- and the second-order necessary conditions are derived beforehand. This problem transformation is adopted from the book [14]. The origin of this transformation is likely traced to the fundamentals of the theory of ordinary differential equations. In this regard, it is also appropriate to mention the so-called vibro-correct solutions [19].

Besides [7], there are some earlier publications available on second-order conditions for impulsive control problems; see, e.g., [8, 13, 21]. In [21], Legendre–Jacobi–Morse-type second-order necessary conditions of optimality for time-optimal control are derived by using, in an essential way, an extremal principle and the notion of index of quasiextremality provided in [1, 2]. The second-order conditions derived in [13] become trivial, that is, degenerate, for abnormal problems, while the ones discussed here, as it has been mentioned, remain informative. In [8], the case of matrix-multiplier $G = G(t)$ was considered. This chapter contains second-order necessary optimality conditions based on the general theory developed in [4].[1]

This chapter is organized as follows. In Sect. 3.2, the problem is stated including the key definitions and hypotheses. In Sect. 3.3, preliminary material is provided including the Frobenius theorem and the description of the main method for investigation. This method suggests a reduction of the original problem to another problem in which matrix G does not depend on x. In Sect. 3.4, by applying the results obtained in Chap. 2, first-order necessary conditions of optimality in the form of the maximum principle are formulated and proved. Then, in Sect. 3.5, the second-order necessary conditions of optimality are addressed by firstly considering the simpler problem with $G = G(t)$, that is, when the matrix-multiplier does not depend on x. For this simpler problem, the required optimality conditions are proved. Finally, in Sect. 3.6 the main result is presented and proved by applying the result of Sect. 3.5. As usual, concluding Sect. 3.7 contains exercises.

3.2 Problem Statement

Consider the following optimal impulsive control problem:

[1] Regarding the issue of abnormal problems, see also review [6].

$$
\begin{array}{ll}
\text{Minimize} & \varphi(p) \\
\text{subject to:} & dx = f(x,u,t)dt + G(x,t)d\mu, \\
& u(t) \in U, \ \text{a.a.}\, t, \\
& \text{range}(\mu) \subset K, \\
& p \in S, \ t \in T.
\end{array}
\tag{3.1}
$$

Here, functions $\varphi : \mathbb{R}^{2n} \to \mathbb{R}^1$, $f : \mathbb{R}^n \times \mathbb{R}^m \times \mathbb{R}^1 \to \mathbb{R}^n$, $G : \mathbb{R}^n \times \mathbb{R}^1 \to \mathbb{R}^{n\times k}$ satisfy certain assumptions specified below in (H1)–(H3), $T = [t_0, t_1]$ is a fixed time interval, $p = (x_0, x_1)$, with $x_0 = x(t_0)$, $x_1 = x(t_1)$, is the endpoint vector, S is a closed set in $\mathbb{R}^{2n}$, U is a closed set in $\mathbb{R}^m$, K is a closed and convex cone in $\mathbb{R}^k$, and $\varphi(p)$ is the cost function supposed to be minimized. As usual, function $u(\cdot)$ in (3.1) stands for conventional control. It has values in the set U and is measurable and essentially bounded w.r.t. Lebesgue measure ℓ. Finally, μ is the impulsive control given by a Borel measure with range in K. Let us state it formally.

Definition 3.1 The impulsive control in Problem (3.1) is defined by Borel measure μ with range in K, that is, $\mu(B) \in K$ for any Borel subset B of T.

The important differences w.r.t. the previous chapters begin with the notion of trajectory. The trajectory will now be defined in a special way in order to be endowed with the property of robustness w.r.t. weak-* topology approximations of measures. For this definition, we need the notion of attached family of trajectories.

Take an impulsive control μ, atom $\tau \in \text{Ds}(\mu)$, and vectors $a \in \mathbb{R}^n, b \in \mathbb{R}^k$. Denote by $z(\cdot) := z(\cdot; \tau, a, b) : [0, 1] \to \mathbb{R}^n$ the solution to the following dynamical system

$$
\begin{cases}
\dot{z}(s) = G(z(s), \tau)b, & s \in [0, 1], \\
z(0) = a,
\end{cases}
$$

which is called *attached system*.

The function of bounded variation $x(\cdot)$ on the interval T is called solution to the differential equation

$$
dx = f(x,u,t)dt + G(x,t)d\mu, \tag{3.2}
$$

corresponding to the control (u, μ) and the starting value x_0, provided that $x(t_0) = x_0$ and, for every $t \in (t_0, t_1]$,

$$
x(t) = x_0 + \int_{t_0}^{t} f(x,u,\varsigma)d\varsigma + \int_{[t_0,t]} G(x,\varsigma)d\mu_c + \sum_{\tau\in \text{Ds}(\mu):\, \tau\le t} \Big(x_\tau(1) - x(\tau^-)\Big), \tag{3.3}
$$

where it is designated: $x_\tau(\cdot) := z(\cdot; \tau, x(\tau^-), \mu(\tau))$, $\mu(\tau) := \mu(\{\tau\})$, and μ_c stands for the continuous component of measure μ. Note that the sum in (3.3) is well defined because the set $\text{Ds}(\mu)$ is at most countable. Moreover, the summation is absolutely convergent due to the definition of z.

This gives rise to the following definition.

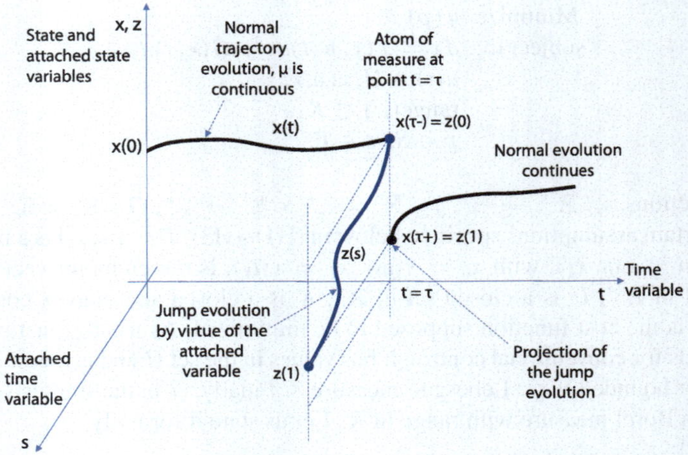

Fig. 3.1 Solution concept (3.3)

Definition 3.2 The trajectory in Problem (3.1) is any function of bounded variation $x(t)$ for which (3.3) holds for some (u, μ) and x_0. The family $\{x_\tau(s)\}_{\tau \in \mathrm{Ds}(\mu)}$ is said to be the family of attached trajectories to $x(t)$, or, in short, attached family. The pair $(x(\cdot), \{x_\tau(\cdot)\}_{\tau \in \mathrm{Ds}(\mu)})$ is said to be the extended trajectory.

This definition is unusual with regard to the conventional types of integration. For example, by using the Lebesgue–Stieltjes integral, it is possible to provide another definition of solution, by replacing (3.3) with

$$\tilde{x}(t) = x_0 + \int_{t_0}^{t} f(\tilde{x}, u, \varsigma)d\varsigma + \int_{[t_0,t]} G(\tilde{x}, \varsigma)d\mu. \qquad (3.4)$$

However, simple examples show that the solution in the sense of (3.4) may fail to exist whenever the measure has atoms (see Exercise 3.1). Moreover, even if solution (3.4) exists, there is always a small arbitrary perturbation of the data (G, μ) such that the solution to the perturbed problem does not exist in the sense of (3.4). Then, solution (3.4) is unsuitable because it is meaningless w.r.t. practical applications. At the same time, the solution in the sense of (3.3) is robust w.r.t. perturbations and, moreover, admits a Cauchy-like existence theorem (see Proposition 3.1). Thus, in spite of the fact that both definitions (3.3) and (3.4) generalize the solution concept of Chap. 1 to the case when G depends on x, it follows that solution (3.3) is well posed while (3.4) is not. Therefore, in what follows, only the solution concept (3.3) is acceptable. The idea for this type of solution is schematically illustrated in Fig. 3.1.

The collection (x, u, μ) is said to be a control process, provided that (3.3) is satisfied. A control process is said to be admissible if the endpoint constraints $p \in S$ are satisfied. An admissible process $(\hat{x}, \hat{u}, \hat{\mu})$ is said to be optimal if, for any admissible process (x, u, μ), the inequality $\varphi(\hat{p}) \leq \varphi(p)$ holds true, where $\hat{p} = (\hat{x}(t_0), \hat{x}(t_1))$. This is the notion of *global minimum*. However, in optimal control theory, specifically with regard to the study of various second-order extremum conditions, it is common to consider weak local types of minima. Therefore, the following definition is also adopted.

Definition 3.3 The admissible process $(\hat{x}, \hat{u}, \hat{\mu})$ is said to be a weakest (finite-dimensional) minimizer in Problem (3.1), provided that for any finite-dimensional subspace $E \subset \mathrm{L}_\infty(T; \mathbb{R}^m) \times \mathrm{C}^*(T; \mathbb{R}^k)$ there exists a number $\varepsilon = \varepsilon(E) > 0$ such that process $(\hat{x}, \hat{u}, \hat{\mu})$ is optimal in Problem (3.1) with additional constraints:

$$|p - \hat{p}| < \varepsilon, \quad \|u - \hat{u}\|_{\mathrm{L}_\infty} + \|\mu - \hat{\mu}\| < \varepsilon, \quad (u, \mu) \in E.$$

The weakest finite-dimensional type of minimum corresponds to the minimum w.r.t. the finite topology of the linear space $\mathscr{X} := \mathbb{R}^n \times \mathrm{L}_\infty(T; \mathbb{R}^m) \times \mathrm{C}^*(T; \mathbb{R}^k)$. In the linear space $\mathscr{X}$, the set is open in the finite topology if its intersection with any finite-dimensional subspace is open, [17]. This topology obviously contains more open subsets than any other topology used in optimal control. This fact justifies the above-given definition.

Overall, both here as well as in the rest of this book, the type of the minima considered is not the priority subject. For an easier perception of the material, one may assume throughout the chapter that the type of minimum is always global. (The global minimum is obviously stronger than the one from Definition 3.3.)

Throughout this chapter, we assume the following hypotheses for the data of the problem to be satisfied:

(H1) The vector-valued function f is twice differentiable w.r.t. x, u for almost all t, f and its partial derivatives w.r.t. x, u up to the second-order are Lebesgue measurable w.r.t. t for all x, u, and, on any bounded set, they are continuous w.r.t. x, u uniformly w.r.t. x, u, t and bounded. The scalar-valued function $\varphi(p)$ is twice, and the matrix-valued function G is thrice continuously differentiable.

(H2) The mappings f, G satisfy the estimate

$$|f(x, u, t)| + |G(x, t)| \leq \mathrm{const} \cdot (1 + |x|) \ \ \forall (x, u, t) \in \mathbb{R}^n \times U \times T.$$

(H3) The following symmetry assumption is imposed on the columns G^i, $i = 1, \dots, k$, of the matrix-valued function G

$$\frac{\partial G^i}{\partial x}(x, t) G^j(x, t) = \frac{\partial G^j}{\partial x}(x, t) G^i(x, t) \ \ \forall i, j, \ \ \forall x \in \mathbb{R}^n, \ \ \forall t \in T, \quad (3.5)$$

which is also known as the property of commutativity of the vector fields G^i or, the *Frobenius condition*.

Let us briefly comment on these hypotheses. Hypothesis (H1) represents some standard assumption imposed on the right-hand side (see, for example, [11]). The higher degree of smoothness of matrix G is dictated by the method of proof. Hypothesis (H2) is the so-called *global solvability condition* under which there exist the global solution $z(\cdot; a, b, \tau)$, and thus, the global solution (3.3) as well. For many cases of interest, such as, for example, necessary optimality conditions discussed below, (H2) can easily be dispensed with by changing f, G smoothly outside of some sufficiently large ball, in such a way that f, G become bounded, and by considering the arguments within this large ball. The commutativity property (3.5) invoked in (H3) is crucial for the forthcoming presentation. Its origin will become clear in the next section.

3.3 Preliminaries

In this section, the main method for investigation based on the Frobenius theorem is described. However, before outlining the details of the method, let us make an observation and adopt a convention. Note that, in (3.1), it is not restrictive to consider that the matrix-valued function G depends only on x, that is, $G = G(x)$. Indeed, the general case is reduced to this one by means of the following simple problem transformation. Introduce a new state variable χ, and consider the new dynamics

$$\begin{aligned} dx &= f(x, u, t)dt + G(x, \chi)d\mu \\ d\chi &= dt, \end{aligned}$$

with $x(t_0) = x_0$, $\chi(t_0) = t_0$. Set $\tilde{f} = (f, 1)$, $\tilde{G} = (G, 0)$ in the sense that the zero row is added to G. It is easy to verify that (H1)–(H3) are valid for f, G if and only if they are valid for $\tilde{f}$, $\tilde{G}$. Moreover, this transformation of the dynamics does not change the trajectory $x(\cdot)$. Therefore, in what follows, the convention is agreed that all the proofs are given for the time-independent case $G = G(x)$, while the main results are stated for the general case $G = G(x, t)$. This convention significantly simplifies the notation.

Let us proceed with the Frobenius theorem (see, for example, in [12, 16]). Consider the following differential equation

$$\xi'(w) = F(w, \xi(w)), \tag{3.6}$$

where $F : \mathbb{R}^n \times \mathbb{R}^k \to \mathbb{R}^{n\times k}$ is some smooth matrix-valued function (having n rows and k columns), $\xi : \mathbb{R}^k \to \mathbb{R}^n$, $w \in \mathbb{R}^k$. Consider also the associated bilinear mapping

$$\mathscr{F}[w_1, w_2] = \mathscr{F}(w, \xi)[w_1, w_2] := \Big(F'_\xi(w, \xi) \circ F(w, \xi) + F'_w(w, \xi)\Big)[w_1, w_2],$$

which is, for each w, ξ, defined over $\mathbb{R}^k \times \mathbb{R}^k$, and takes on values in $\mathbb{R}^n$. (Above, the symbol $\circ$, as usual, stands for the composition). Thus, $\mathscr{F} = (\mathscr{F}_1, \mathscr{F}_2, \dots, \mathscr{F}_n)$, where

$$\mathscr{F}_i[w_1, w_2] = \left\langle \left(F'^*_{\xi^i}(w, \xi)F(w, \xi) + (F^i)'^*_w(w, \xi) \right) w_1, w_2 \right\rangle, \quad i = 1, 2, \ldots, n,$$

being F^i the rows of F. (Above, F'_{ξ^i} is the elementwise derivative matrix of F w.r.t. ξ^i, and $(F^i)'^*_w$ is the Jacobi matrix.)

Theorem 3.1 (Frobenius) *Suppose that the bilinear mapping $\mathscr{F}(w, \xi)$ is symmetric for all w, ξ in some neighborhood of a given point (w_0, ξ_0). Symmetry stands for*

$$\mathscr{F}[w_1, w_2] = \mathscr{F}[w_2, w_1] \ \forall w_1, w_2.$$

Then, there exists a local unique solution $\xi(w)$ to Eq. (3.6) in the proximity of point w_0, such that $\xi(w_0) = \xi_0$. Furthermore, if the symmetry property holds for all $w \in \mathbb{R}^k$ and $\xi \in \mathbb{R}^n$, and if the global solvability condition

$$|F(w, \xi)| \leq const \cdot (1 + |\xi|) \ \forall w \in \mathbb{R}^k, \quad \forall \xi \in \mathbb{R}^n,$$

is satisfied, then this solution exists everywhere on $\mathbb{R}^k$ and is unique.

Note that when $k = 1$, (3.6) represents a system of ordinary differential equations, the symmetry automatically being satisfied. Then, the Frobenius theorem is just the Cauchy existence theorem. However, when $k > 1$, (3.6) is a system of partial differential equations to which this existence theorem, in general, is not applicable. At the same time, the Frobenius theorem asserts that the Cauchy-like results are still valid, provided that $\mathscr{F}(w, \xi)[\cdot, \cdot]$ is a symmetric bilinear map for all w, ξ from the considered domain. This property of symmetry is also called Frobenius condition and, as is ensured below, it follows to be the commutativity property (3.5), if $F = G$. This sheds light on the origin of the term "Frobenius condition" used for (3.5). It appears that this condition is the weakest possible assumption under which an important set of results, asserted in the theory of ordinary differential equations, is still relevant in the context of Eq. (3.6).

The Frobenius theorem is invertible. That is, if solution to (3.6) exists, then the bilinear mapping $\mathscr{F}$ is symmetric. So, the Frobenius theorem provides necessary and sufficient condition of solvability. At the same time, the necessary condition (that is, the invertibility) is a simple exercise to prove (see Exercise 3.2). The sufficient condition, that is, the solvability itself, requires more effort—the proof is reduced to application of the classical existence theorem for ordinary differential equations along the rays emitted from w_0. Then, due to the Frobenius condition, it suffices to ensure differentiability of the resulting function over $\mathbb{R}^k$.

The assertion of the Frobenius theorem can be extended as follows. Let the global solvability condition be satisfied. Let us perturb the point ξ_0 and solve the Eq. (3.6) with the initial condition $\xi(w_0) = y$, where $y \in \mathbb{R}^n$ is a parameter. Then, ξ is a function of w and of the initial data y, $\xi = \xi(w, y)$. It follows that, just as in the case of $k = 1$, i.e., when $w = t \in \mathbb{R}^1$, the map ξ w.r.t. y is from the same class of smoothness as F w.r.t. ξ. Moreover, the Jacobian ξ_y' is nonsingular everywhere. This full version of the Frobenius theorem will be used further on.[2]

Let us proceed with the main method of investigation employed in this chapter. Consider matrix $G = G(x)$ from problem formulation (3.1), and the differential system

$$\begin{cases} \xi_w'(w, y) = G(\xi(w, y)) \\ \xi(0, y) = y, \end{cases}$$

where $w \in \mathbb{R}^k$, $y \in \mathbb{R}^n$, $\xi : \mathbb{R}^k \times \mathbb{R}^n \to \mathbb{R}^n$. Let us examine the bilinear map $\mathcal{F}$ defined above and decode the symmetry condition when $F = G$. Assume, for simplicity, $n = 1$. Then, G is a row vector. Bilinear mapping $\mathcal{F} = \mathcal{F}(x)$ over $\mathbb{R}^k \times \mathbb{R}^k$, $x \in \mathbb{R}^1$, equals

$$\mathcal{F}(x) = G'^{*}(x)G(x) = \begin{bmatrix} G^{1'}(x) \\ G^{2'}(x) \\ \dots \\ G^{k'}(x) \end{bmatrix} \Big[G^1(x), G^2(x), \dots, G^k(x) \Big] =$$

$$= \begin{bmatrix} G^{1'}(x)G^1(x) & G^{1'}(x)G^2(x) & \cdots & G^{1'}(x)G^k(x) \\ G^{2'}(x)G^1(x) & G^{2'}(x)G^2(x) & \cdots & G^{2'}(x)G^k(x) \\ \dots & \dots & \dots & \dots \\ G^{k'}(x)G^1(x) & G^{k'}(x)G^2(x) & \cdots & G^{k'}(x)G^k(x) \end{bmatrix}.$$

Then, the symmetry implies that

$$G^{i'}(x)G^j(x) = G^{j'}(x)G^i(x) \;\; \forall i, j = 1, \dots, k,$$

and thus, (H3) is obtained. For $n > 1$, the arguments are by verbatim repetition. In this regard, see also Exercise 3.3.

Therefore, in view of (H2), (H3), and the Frobenius theorem, the unique solution $\xi(w, y)$ exists everywhere on $\mathbb{R}^k \times \mathbb{R}^n$ having the same class of smoothness as G, that is, thrice continuously differentiable. Moreover, the Jacobian ξ_y' is nonsingular everywhere. Take any points $w_0 \in \mathbb{R}^k$, $y_0 \in \mathbb{R}^n$ and consider the equation

$$x = \xi(w, y)$$

in the proximity of (w_0, y_0). By applying the implicit function theorem, there exists a function $\eta(w, x)$ with values in $\mathbb{R}^n$ such that

$$x = \xi(w, \eta(w, x)) \tag{3.7}$$

in some neighborhood of the point (w_0, x_0), where $x_0 = \xi(w_0, y_0)$. Function η has the same class of smoothness as ξ. By differentiating (3.7) w.r.t. x,

[2] Regarding the Frobenius theorem, the reader can find more details in, e.g., [12, 16].

$$I = \xi_y'(w, \eta(w, x))\eta_x'(w, x), \tag{3.8}$$

where I is the identity $n \times n$ matrix. By differentiating (3.7) w.r.t. w-variable, we have

$$0 = \xi_w'(w, \eta(w, x)) + \xi_y'(w, \eta(w, x))\eta_w'(w, x).$$

By using nonsingularity of the Jacobian ξ_y' and by multiplying this equality by the inverse matrix $[\xi_y']^{-1}$, in view of (3.8), the equation for η is derived:

$$\eta_x'(w, x)G(x) + \eta_w'(w, x) = 0, \quad \eta(w_0, x_0) = y_0.$$

This system of equations is called conjugate to the system $\xi_w' = G(\xi)$. The solution η is unique in the above-mentioned neighborhood. At the same time, for each w, $\xi(w, \mathbb{R}^n) = \mathbb{R}^n$, as follows from the continuity of solution w.r.t. the initial data. This allows us to repeat the construction for an arbitrary point (w_0, x_0) and, thereby, to ensure the global existence of solution η to the conjugate system

$$\begin{cases} \eta_x'(w, x)G(x) + \eta_w'(w, x) = 0 \\ \eta(0, x) = x. \end{cases}$$

Should we consider, over some time interval, the differential system $\dot{x} = G(x)m$, $\dot{w} = m$, where m is some integrable function, function η becomes the first integral, as $\eta(x(t), w(t)) \equiv \text{const}$. Therefore, it is also the first integral for the attached differential system $\dot{z}(s) = G(z(s))b$ with $w(s) = b \cdot s$, $s \in [0, 1]$.

All these arguments are invertible, and by applying the implicit function theorem as has been done previously, one can pass from η to ξ. Therefore, the following relations, similar to (3.7), (3.8), but w.r.t. y also hold:

$$\eta(w, \xi(w, y)) = y \quad \Rightarrow \quad I = \eta_x'(w, \xi(w, y))\xi_y'(w, y).$$

These will also be used further on.

Now, along with (3.2), consider the following reduced dynamics:

$$\dot{y} = \eta_x'(w, \xi(w, y)) f(\xi(w, y), u, t), \tag{3.9}$$

where $w = w(t)$ is the distribution function of μ, that is, (F here should not be confused with F from the Frobenius theorem)

$$w(t) := F(t; \mu), \quad t \in T.$$

The idea of the method is as follows. The differential control systems (3.2) and (3.9) are equivalent, as there exists a one-to-one correspondence between their trajectories, while the reachable sets are diffeomorphic. At the same time, Eq. (3.9) is of conventional type, and $dw = d\mu$. This means that the above-mentioned difficulty of x-dependency is overcome. The following lemma constitutes this idea.

Lemma 3.1 *The trajectory $x(t)$ of system (3.2) associated with the controls u, μ and the initial value $x(t_0) = a$, and the trajectory $y(t)$ of system (3.9) generated by the same u, μ and $y(t_0) = a$, are, when G does not depend on t, i.e., $G = G(x)$, related as follows:*

$$x(t) = \xi(w(t), y(t)), \quad y(t) = \eta(w(t), x(t)), \quad t \in T. \tag{3.10}$$

The proof of this lemma is preceded with the following important assertion (existence of solution and its robustness in the sense of weak-* approximation).

Proposition 3.1 *Solution $x(t)$ of Eq. (3.2) satisfying (3.3) w.r.t. the controls u, μ and the initial value $x(t_0) = x_0$ exists and is unique.*

Moreover, let $\{\mu_i\}$ be a given sequence of absolutely continuous measures such that $\mu_i \overset{w*}{\to} \mu$, $|\mu_i| \overset{w*}{\to} \nu$. Denote $m_i(t) = \frac{d\mu_i}{dt}$. Let $x_i(\cdot)$ be such an absolutely continuous function that

$$\dot{x}_i(t) = f(x_i(t), u(t), t) + G(x_i(t), t)m_i(t), \quad x_i(t_0) = x_0,$$

for a.a. $t \in [t_0, t_1]$.

Then,

$$x_i(t) \to x(t) \ \ \forall t \in Cs(\nu). \tag{3.11}$$

Proof First, suppose that $f = 0$ and that the map G does not depend on time, that is, $G = G(x)$. Then, the equation for x_i is

$$\dot{x}_i(t) = G(x_i(t))m_i(t).$$

Since m_i is integrable on T, in view of (H2), the solution x_i with $x_i(t_0) = x_0$ exists for all i and, over the entire time interval T, see, e.g., Chap. 8 in [15]. Functions m_i are uniformly bounded in the L_1-norm due to the uniform boundedness principle and weak-* convergence of the measures. Then, in view of the Gronwall inequality—verify Exercise 3.4—trajectories x_i are uniformly bounded in the C-norm. Then, in view of Exercise 1.3, they are also uniformly bounded in the BV-norm, i.e., $\mathrm{Var}\, x_i|_T \le$ const. By applying the Helly theorem, it is deduced that there exists a function of bounded variation $\tilde{x}$ such that, after passing to a subsequence, $x_i(t) \to \tilde{x}(t)\ \forall t \in T$.

This is a straightforward task to ensure the existence of decomposition

$$\mu_i = \mu_{c,i} + \mu_{d,i},$$

where $\mu_{c,i} \overset{w*}{\to} \mu_c$, $\mu_{d,i} \overset{w*}{\to} \mu_d$, $\mu_{c,i}$, $\mu_{d,i}$ are absolutely continuous, and μ_c, μ_d are the continuous and discrete (atomic) components of μ. Moreover, $|\mu_{c,i}| \overset{w*}{\to} \nu_c$, $|\mu_{d,i}| \overset{w*}{\to} \nu_d$.

Consider

$$x_i(t) = x_0 + \int_{t_0}^{t} G(x_i(\varsigma))d\mu_i = x_0 + \int_{t_0}^{t} G(x_i(\varsigma))d(\mu_i - \mu_{\mathrm{d},i}) + \int_{t_0}^{t} G(x_i(\varsigma))d\mu_{\mathrm{d},i}.$$

By passing to the limit here, in view of Exercise 3.5 (Lemma 3.1 in [18]), we have

$$\tilde{x}(t) = x_0 + \int_{t_0}^{t} G(\tilde{x}(\varsigma))d\mu_{\mathrm{c}} + A(t),$$

where $A(t) := \lim_{i\to\infty} \int_{t_0}^{t} G(x_i(\varsigma))d\mu_{\mathrm{d},i}$. Note that A is a function of bounded variation which, by construction, is continuous on $\mathrm{Cs}(\nu)\cap(t_0, t_1)$. (This follows from the above estimates and regularity of Borel measure.) Then, $\tilde{x}$ has the same property.

Take any $\theta \in \mathrm{Cs}(\nu)$, $\theta > t_0$. Let us show that

$$A(\theta) = \sum_{\tau\le\theta}\Big(z(1; \tilde{x}(\tau^-), \mu(\tau), \tau) - \tilde{x}(\tau^-)\Big). \tag{3.12}$$

Consider any atom $\tau \in \mathrm{Ds}(\nu)$: $\tau \le \theta$. There exist sequences $\{\tau_i^+\}$ and $\{\tau_i^-\}$ such that both converge to τ, and $|\mu_{\mathrm{d},i}|(\Delta_i) \to \nu(\tau)$ as $i \to \infty$, where $\Delta_i := [\tau_i^-, \tau_i^+]$. From the above considerations, $x_i(\tau_i^-) \to \tilde{x}(\tau^-)$, $x_i(\tau_i^+) \to \tilde{x}(\tau^+)$.

From the properties of ξ, η, we have, for all $t \in T$,

$$x_i(t) = \xi(w_i(t), \eta(w_i(t), x_i(t))),$$

where $w_i(t) := F(t; \mu_i)$. As was noted earlier, η is the first integral for the differential system $\dot{x} = G(x)m$, $\dot{w} = m$, where $m(\cdot)$ is any integrable function. Then,

$$\eta(w_i(\tau_i^-), x_i(\tau_i^-)) = \eta(w_i(\tau_i^+), x_i(\tau_i^+)).$$

Hence,

$$x_i(\tau_i^+) = \xi(w_i(\tau_i^+), \eta(w_i(\tau^-), x_i(\tau_i^-))). \tag{3.13}$$

Let us replace the function w_i with the function $\bar{w}_i$ defined on the time interval Δ_i as follows

$$\bar{w}_i(t) := w_i(\tau_i^-) + \frac{t - \tau_i^-}{\ell(\Delta_i)}(w_i(\tau_i^+) - w_i(\tau_i^-)),$$

and set $\bar{x}_i(t) := \xi(\bar{w}_i(t), \eta(w_i(\tau_i^-), x_i(\tau_i^-)))$, $t \in \Delta_i$. It is clear that $\bar{w}_i(\tau_i^-) = w_i(\tau_i^-)$, $\bar{w}_i(\tau_i^+) = w_i(\tau_i^+)$. Then, due to (3.13), $\bar{x}_i(\tau_i^-) = x_i(\tau_i^-)$, $\bar{x}_i(\tau_i^+) = x_i(\tau_i^+)$. By differentiating w.r.t. t on Δ_i, we obtain that

$$\begin{cases} \dot{\bar{x}}_i(t) = G(\bar{x}_i(t))\frac{w_i(\tau_i^+)-w_i(\tau_i^-)}{\ell(\Delta_i)} \\ \bar{x}_i(\tau_i^-) = x_i(\tau_i^-). \end{cases}$$

Thus, on Δ_i, the original differential system was substituted by a system with linear (Lebesgue) measure. By applying the obvious linear transformation, we proceed to the attached system and deduce that

$$x_i(\tau_i^+) = z(1; \tau, x_i(\tau_i^-), w_i(\tau_i^+) - w_i(\tau_i^-)).$$

By passing to the limit here, which, under (H2), is straightforward,

$$\tilde{x}(\tau^+) = z(1; \tau, \tilde{x}(\tau^-), \mu(\tau)).$$

Thus, (3.12) is proved.

Let $x \in \mathrm{BV}(T; \mathbb{R}^n)$ be the right-continuous function on (t_0, t_1) such that $x(t) = \tilde{x}(t)$ a.e., and $x(t_0) = \tilde{x}(t_0)$, $x(t_1) = \tilde{x}(t_1)$. In view of the above arguments, it has just been proven that $x(\cdot)$ satisfies (3.3). Thus, the existence of solution is justified. As it can be concluded by Exercise 3.6, the solution is unique. Moreover, from the above arguments, (3.11) clearly follows, with regard to selection of a subsequence. Since these arguments can be applied to any subsequence of the original sequence of trajectories, (3.11) is true for the entire sequence $\{x_i\}$.

When $f \neq 0$, the arguments are the same due to the absolute continuity of the Lebesgue integral. The case $G = G(x, t)$ is treated according to the reduction discussed earlier in this section. □

Proposition 3.1 ensures the well-posedness, or robustness, of the solution to Eq. (3.2) in the sense given by (3.3). In Chap. 4, with the help of a more powerful technique, this assertion will be significantly generalized, as the robustness will be ensured in extended sense for impulsive control systems which do not satisfy the Frobenius condition. Let us proceed to the proof of Lemma 3.1.

Proof of Lemma 3.1. First, let us prove (3.10) for the case of absolutely continuous μ. Assume that $d\mu = m(t)dt$, so $\dot{w}(t) = m(t)$. Consider any feasible trajectory $y(\cdot)$ satisfying (3.9) w.r.t. u, μ. Let us ensure that the function

$$x(t) = \xi(w(t), y(t)), \quad t \in T$$

satisfies (3.3) w.r.t. u, μ and $x(t_0) = y(t_0)$. Since $w(t_0) = 0$, the relation $x(t_0) = y(t_0)$ follows from the definition of ξ. By differentiating $x(t)$, we have

$$\begin{aligned} &\dot{x}(t) = \xi'_w(w(t), y(t))\dot{w}(t) + \xi'_y(w(t), y(t))\dot{y}(t) = \\ &G(\xi(w(t), y(t)))m(t) + \xi'_y(w(t), y(t))\eta'_x(w(t), \xi(w(t), y(t)))f(\xi(w(t), y(t)), u(t), t) = \\ &G(x(t))m(t) + I \cdot f(x(t), u(t), t). \end{aligned}$$

Then, $x(\cdot)$ satisfies (3.3) w.r.t. u, μ, $y(t_0)$. Above, (3.9) was used as well as the properties of ξ, η derived earlier.

Conversely, take $x(\cdot)$ satisfying (3.3) w.r.t. u, μ. Consider the function

$$y(t) = \eta(w(t), x(t)), \quad t \in T.$$

By differentiating, we have

$$\begin{aligned}
&\dot{y}(t) = \eta'_w(w(t), x(t))\dot{w}(t) + \eta'_x(w(t), x(t))\dot{x}(t) = \\
&\eta'_w(w(t), x(t))m(t) + \eta'_x(w(t), x(t))\Big(f(x(t), u(t), t) + G(x(t))m(t)\Big) = \\
&\Big(\eta'_w(w(t), x(t)) + \eta'_x(w(t), x(t))G(x(t))\Big)m(t) + \eta'_x(w(t), x(t))f(x(t), u(t), t) = \\
&\eta'_x(w(t), x(t))f(x(t), u(t), t) = \\
&\eta'_x(w(t), \xi(w(t), \eta(w(t), x(t))))f(\xi(w(t), \eta(w(t), x(t))), u(t), t) = \\
&\eta'_x(w(t), \xi(w(t), y(t)))f(\xi(w(t), y(t)), u(t), t).
\end{aligned}$$

This means that $y(\cdot)$ satisfies (3.9). Moreover, $y(t_0) = x(t_0)$. Above, the properties of ξ, η have again been used, in particular, the properties of the conjugate system and (3.7). Therefore, (3.10) is proved for the absolute continuous measures.

Now, let us prove the assertion in the general case. Consider $x(t)$ to be the solution to (3.2). Consider any weak-* approximation of μ by absolute continuous measures μ_i such that $\mu_i \overset{w*}{\to} \mu$, and $d\mu_i = m_i(t)dt$. In view of the robustness property stated in Proposition 3.1, trajectories $x_i(\cdot)$ satisfying (3.3) w.r.t. u, μ_i and $x_i(t_0) = x(t_0)$ exist and $x_i(t) \to x(t)$ for a.a. t and for $t = t_1$. In view of what has been already proven, the functions

$$y_i(t) = \eta(w_i(t), x_i(t)),$$

where $w_i(t) := F(t; \mu_i)$, satisfy (3.9), and, due to weak-* convergence and (3.11), are uniformly bounded: $\|y_i\|_C \leq \text{const}\ \forall\, i$. Then, since $w_i(t) \to w(t) = F(t; \mu)$ for a.a. t and for $t = t_1$, it is a straightforward task to ensure that $y_i \rightrightarrows y \in C(T; \mathbb{R}^n)$, where $y(\cdot)$ is the solution to (3.9) w.r.t. u, μ and $y(t_0) = x(t_0)$. At the same time, it is clear that $y(t) = \eta(w(t), x(t))$, as $w(\cdot)$ and $x(\cdot)$ are right-continuous on (t_0, t_1).

Conversely, consider $y(t)$ to be the solution to (3.9). Consider any weak-* approximation of μ by absolute continuous measures μ_i: $\mu_i \overset{w*}{\to} \mu$, with $d\mu_i = m_i(t)dt$. In view of (H1), trajectories $y_i(\cdot)$ satisfying (3.9) w.r.t. u, μ_i and $y_i(t_0) = y_0$ exist for all sufficiently large i, and $y_i(t) \rightrightarrows y(t)$. In view of what has been already proven, the functions

$$x_i(t) = \xi(w_i(t), y_i(t))$$

satisfy (3.3) w.r.t. u, μ_i and $x_i(t_0) = y(t_0)$ and due to (3.11) converge to the solution $x(t)$ of (3.2) w.r.t. u, μ and $x(t_0) = y(t_0)$. It is clear that $x(t) = \xi(w(t), y(t))$. □

3.4 Maximum Principle

From henceforth, the first- and the second-order necessary optimality conditions for the impulsive control problem (3.1) under the Frobenius condition are investigated. In this section, the first-order conditions in the form of the maximum principle are derived. The idea behind the proof is based on a reduction to a simpler problem, in which the x-dependency of the matrix-multiplier G of the control measure is removed. This reduction, in turn, is based on Lemma 3.1.

Denote by H the Hamilton–Pontryagin function

$$H(x, u, \psi, t) := \langle f(x, u, t), \psi \rangle,$$

and by Q the following vector-valued function $Q(x, \psi, t) := G^*(x, t)\psi$.

Theorem 3.2 *Let $(\hat{x}, \hat{u}, \hat{\mu})$ be an optimal process in Problem (3.1). Then, there exist a number $\lambda \geq 0$ and a vector-valued function of bounded variation ψ such that*

$$\lambda + |\psi(t)| \neq 0 \;\; \forall t \in T, \tag{3.14}$$

$$\psi(t) = \psi(t_0) - \int_{t_0}^{t} \frac{\partial H}{\partial x}(\varsigma)d\varsigma - \int_{[t_0,t]} \frac{\partial}{\partial x}\Big\langle Q(\varsigma), d\hat{\mu}_c \Big\rangle + \sum_{\tau \in Ds(\hat{\mu}):\, \tau \leq t} \Big(\psi_\tau(1) - \psi_\tau(0)\Big) \;\; \forall t \in (t_0, t_1], \tag{3.15}$$

$$(\psi(t_0), -\psi(t_1)) \in \lambda \varphi'(\hat{p}) + N_S(\hat{p}), \tag{3.16}$$

$$\max_{u \in U} H(u, t) = H(t) \;\; a.e. \tag{3.17}$$

$$\langle Q(t), v \rangle \leq 0 \;\; \forall v \in K, \;\; \forall t \in T, \tag{3.18}$$

$$\int_T \langle Q(t), d\hat{\mu} \rangle = 0, \tag{3.19}$$

where function ψ_τ, for each $\tau \in Ds(\hat{\mu})$, is the solution to the equation:

$$\begin{cases} \dot{\psi}_\tau(s) = -\frac{\partial}{\partial x}\langle Q(\hat{x}_\tau(s), \psi_\tau(s), \tau), \hat{\mu}(\tau) \rangle, \\ \psi_\tau(0) = \psi(\tau^-), \;\; s \in [0, 1]. \end{cases}$$

Moreover, the functions $Q^j(t)$ are absolutely (and even Lipschitz) continuous on T, for $j = 1, \ldots, k$.

Here, $\hat{p} = (\hat{x}(t_0), \hat{x}(t_1))$. As usual, if some of the arguments of H, Q and their partial derivatives are omitted, then it means that the extremal values substitute the missing arguments, i.e., $H(t) = H(\hat{x}(t), \hat{u}(t), t, \psi(t))$, or $H(u,t) = H(\hat{x}(t), u, t, \psi(t))$.

Let us briefly comment on the formulated maximum principle. Condition (3.15) is the equation for the adjoint function $\psi(\cdot)$ which, unlike (1.11), is now a function of bounded variation, just as $x(\cdot)$, and, therefore, may have jumps. This is due to the fact that G depends on x. The jumps of ψ are computed via solutions ψ_τ of the adjoint conjugate system. The transversality condition and the maximum condition residing in (3.16) and (3.17), respectively, are exactly the same as those in preceding chapters. Conditions (3.18) and (3.19) represent the impulsive maximum condition. Condition (3.18) means that $Q(t) \in K^\circ\ \forall t$, where K° is the polar cone. When $K = \mathbb{R}^k_+$, they become (1.14) and (1.15), respectively, as $Q(t)$ is continuous. Note that, under (H3), the absolute continuity of $Q(t)$ is not an independent condition, but follows directly from the properties of the Hamiltonian system (3.3), (3.15); see Exercise 3.7.

Proof Let $G = G(x)$. Along with the original problem (3.1), consider the following reduced impulsive control problem.

$$
\begin{aligned}
\text{Minimize}\quad & \varphi(y_0, \zeta_1)\\
\text{subject to:}\quad & dy = g(w, y, u, t)dt,\\
& dw = d\mu,\\
& u(t) \in U, \ \text{ a.a. } t,\\
& \text{range}(\mu) \subset K, \ t \in T,\\
& w_0 = 0, \ \zeta_1 = \xi(w_1, y_1), \ (y_0, \zeta_1) \in S.
\end{aligned}
\tag{3.20}
$$

Here, the function g is defined by

$$
g(w, y, u, t) := \eta'_x(w, \xi(w, y)) f(\xi(w, y), u, t),
$$

$y_0 = y(t_0)$, $y_1 = y(t_1)$, $w_0 = w(t_0)$, $w_1 = w(t_1)$, and $\zeta_1 = \zeta(t_1)$, where ζ is an auxiliary state variable such that $d\zeta = 0$.

The two Problems (3.1) and (3.20) are equivalent. This assertion readily follows from Lemma 3.1, as, in view of (3.10), the admissible processes of the considered problems are connected by the following one-to-one correspondence:

$$
(x, u, \mu) \rightarrowtail (y = \eta(w, x), w, u, \mu), \quad \text{and} \quad (y, w, u, \mu) \rightarrowtail (x = \xi(w, y), u, \mu),
$$

which obviously preserves the value of the cost, as well as the endpoint constraint.

Let the process $(\hat{x}, \hat{u}, \hat{\mu})$ be optimal in (3.1). Then, the process $(\hat{y} = \eta(\hat{w}, \hat{x}), \hat{w}, \hat{u}, \hat{\mu})$, where $\hat{w}(t) = F(t; \hat{\mu})$, is optimal in (3.20). Let us apply the maximum principle of Chap. 2. The functions H, Q take the following form

$$
\tilde{H}(w, y, u, \psi_y, \psi_w, t) := \langle \psi_y, g(w, y, u, t) \rangle, \quad \tilde{Q}(\psi_w) := \psi_w.
$$

Then, there exist a number $\lambda \geq 0$, a vector $a \in \mathbb{R}^n$, and absolutely continuous row vector functions ψ_y, ψ_w, ψ_ζ which do not vanish simultaneously and such that

$$\begin{aligned}\dot{\psi}_y(t) &= -\tilde{H}'_y(t),\\ \dot{\psi}_w(t) &= -\tilde{H}'_w(t),\\ \dot{\psi}_\zeta(t) &= 0, \ \ t \in T,\end{aligned} \tag{3.21}$$

$$\begin{aligned}&(\psi_y(t_0), \psi_w(t_0), \psi_\zeta(t_0), -\psi_y(t_1), -\psi_w(t_1), -\psi_\zeta(t_1)) \in \\ &\lambda\Big(\varphi'_{x_0}(\hat{y}_0, \hat{\zeta}_1), 0, 0, 0, 0, \varphi'_{x_1}(\hat{y}_0, \hat{\zeta}_1)\Big) + \\ &\Big(0, 0, 0, -a\xi'_y(\hat{w}_1, \hat{y}_1), -a\xi'_w(\hat{w}_1, \hat{y}_1), a\Big) + N_{\bar{S}}(\hat{p}^*),\end{aligned} \tag{3.22}$$

$$\max_{u \in U} \tilde{H}(u, t) = \tilde{H}(t) \ \text{ a.e.}, \tag{3.23}$$

$$\langle \psi_w(t), v \rangle \le 0 \ \ \forall\, v \in K, \ \ \forall\, t \in T, \tag{3.24}$$

$$\int_T \langle \psi_w(t), d\hat{\mu} \rangle = 0, \tag{3.25}$$

where $\hat{p}^* := (\hat{y}_0, \hat{w}_0, \hat{\zeta}_0, \hat{y}_1, \hat{w}_1, \hat{\zeta}_1)$, and

$$\bar{S} := \{p^* = (y_0, w_0, \zeta_0, y_1, w_1, \zeta_1) : w_0 = 0, (y_0, \zeta_1) \in S\}.$$

Since $N_{\bar{S}}(\hat{p}^*) \subseteq \Pi = \{p^* : \zeta_0 = 0, y_1 = 0, w_1 = 0\}$, and due to the fact that ψ_ζ is constant, readily deduce from (3.22) that $\psi_\zeta = 0$, and, then, also

$$a = \psi_y(t_1)[\xi'_y(\hat{w}_1, \hat{y}_1)]^{-1},$$

$$(\psi_y(t_0), -\psi_y(t_1)[\xi'_y(\hat{w}_1, \hat{y}_1)]^{-1}) \in \lambda\varphi'(\hat{y}_0, \hat{\zeta}_1) + N_S(\hat{p}), \tag{3.26}$$

$$\psi_w(t_1) = \psi_y(t_1)[\xi'_y(\hat{w}_1, \hat{y}_1)]^{-1}\xi'_w(\hat{w}_1, \hat{y}_1). \tag{3.27}$$

A thorough analysis of the conditions obtained above suggests the following function as an "appropriate candidate" to satisfy Theorem 3.1 (as row vector):

$$\psi(t) := \psi_y(t)\eta'_x(\hat{w}(t), \hat{x}(t)) = \psi_y(t)\eta'_x(\hat{w}(t), \xi(\hat{w}(t), \hat{y}(t))) = \psi_y(t)[\xi'_y(\hat{w}(t), \hat{y}(t))]^{-1}.$$

The set of multipliers λ, ψ obviously satisfy (3.16), (3.17). It immediately follows from the definition of ψ, and from (3.26) and (3.23), respectively. Note that for deriving the transversality condition (3.16), the equality $\xi'_y(\hat{w}_0, \hat{y}_0) = I$ is used.

Let us prove (3.15). First, assume that $\hat{\mu}$ is absolutely continuous. Let $\frac{d\hat{\mu}}{dt} = \hat{m}(t)$. By differentiating ψ,

$$\begin{aligned}
&\dot{\psi}(t) = \tfrac{d}{dt}(\psi_y(t)\eta'_x(\hat{w}(t), \hat{x}(t))) = \dot{\psi}_y(t)\eta'_x(\hat{w}(t), \hat{x}(t)) + \psi_y(t)\tfrac{d}{dt}(\eta'_x(\hat{w}(t), \hat{x}(t))) = \\
&-\psi_y(t) \cdot \Big[\eta''_{xx}(\hat{w}(t), \hat{x}(t)) f(\hat{x}(t), \hat{u}(t), t)\xi'_y(\hat{w}(t), \hat{y}(t)) + \\
&\eta'_x(\hat{w}(t), \hat{x}(t)) f'_x(\hat{x}(t), \hat{u}(t), t)\xi'_y(\hat{w}(t), \hat{y}(t))\Big]\eta'_x(\hat{w}(t), \hat{x}(t)) + \\
&\psi_y(t)\Big[\eta''_{xw}(\hat{w}(t), \hat{x}(t))\hat{m}(t) + \eta''_{xx}(\hat{w}(t), \hat{x}(t))(f(\hat{x}(t), \hat{u}(t), t) + G(\hat{x}(t))\hat{m}(t))\Big] = \\
&-\psi(t) f'_x(\hat{x}(t), \hat{u}(t), t) + \psi_y(t)\Big[\eta''_{xw}(\hat{w}(t), \hat{x}(t)) + \eta''_{xx}(\hat{w}(t), \hat{x}(t))G(\hat{x}(t))\Big]\hat{m}(t).
\end{aligned}$$

Above, (3.21), relations $\hat{x}(t) = \xi(\hat{w}(t), \hat{y}(t))$, $\hat{y}(t) = \eta(\hat{w}(t), \hat{x}(t))$, the property (3.8), and the symmetry property of the second-order derivative have been used.

From the definition of η, it easily follows that

$$\eta''_{xx}(w, x)G(x) + \eta'_x(w, x)G'_x(x) + \eta''_{wx}(w, x) = 0, \tag{3.28}$$

whence it can be concluded that

$$\dot{\psi}(t) = -\psi(t) f'_x(\hat{x}(t), \hat{u}(t), t) - \psi(t)G'_x(\hat{x}(t))\hat{m}(t) = -H'_x(t) - Q'_x(t)\hat{m}(t).$$

So, (3.15) is proved for absolutely continuous measures. If $\hat{\mu}$ is an arbitrary measure, this equation still holds true as $\hat{\mu}$ can be weakly-* approximated by absolutely continuous measures. Then, (3.15) is derived similarly to the proof of Proposition 3.1. Note that, in view of the formula $\psi(t) = \psi_y(t)\eta'_x(\hat{w}(t), \hat{x}(t))$, the jumps of ψ do not depend on the method of approximation by absolutely continuous measures (see Exercise 3.8).

Let us prove that

$$Q(t) = \psi_w(t) \quad \forall t \in T. \tag{3.29}$$

Then, (3.18) and (3.19) are proved in view of (3.24), (3.25).

Note that at point $t = t_1$, Equality (3.29) is already proved due to (3.27), as $\xi'_w(\hat{w}_1, \hat{y}_1) = G(\hat{x}(t_1))$. Thus, if $\hat{\mu}$ is absolutely continuous, it is sufficient to show that the derivatives of $Q(t)$, $\psi_w(t)$ w.r.t. t coincide, for a.a. $t \in T$. By differentiating these functions,

$$\begin{aligned}
&\dot{\psi}_w(t) = -\psi_y(t)\Big[(\eta''_{xw}(\hat{w}(t), \hat{x}(t)) + \eta''_{xx}(\hat{w}(t), \hat{x}(t))G(\hat{x}(t))) f(\hat{x}(t), \hat{u}(t), t) + \\
&\eta'_x(\hat{w}(t), \hat{x}(t)) f'_x(\hat{x}(t), \hat{u}(t), t)G(\hat{x}(t))\Big] = \\
&\psi(t)\Big[G'_x(\hat{x}(t)) f(\hat{x}(t), \hat{u}(t), t) - f'_x(\hat{x}(t), \hat{u}(t), t)G(\hat{x}(t))\Big];
\end{aligned}$$

$$\begin{aligned}
&\dot{Q}(t) = \tfrac{d}{dt}[\psi(t)G(\hat{x}(t))] = -\Big(\psi(t) f'_x(\hat{x}(t), \hat{u}(t), t) + \psi(t)G'_x(\hat{x}(t))\hat{m}(t)\Big)G(\hat{x}(t)) + \\
&\psi(t)G'_x(\hat{x}(t))\Big(f(\hat{x}(t), \hat{u}(t), t) + G(\hat{x}(t))\hat{m}(t)\Big) = \\
&\psi(t)\Big[G'_x(\hat{x}(t)) f(\hat{x}(t), \hat{u}(t), t) - f'_x(\hat{x}(t), \hat{u}(t), t)G(\hat{x}(t))\Big].
\end{aligned}$$

Here, (3.15), (3.21), (3.28), and the Frobenius condition have been used. (See also Exercise 3.7.) If $\hat{\mu}$ is arbitrary, (3.29) is true in view of the approximation by absolutely continuous measures and by virtue of Proposition 3.1; see Exercise 3.8. Thus, (3.29) and therefore (3.18), (3.19) are proved. Note that the map $Q(t)$ is Lipschitz continuous, as a consequence of (3.29).

The nontriviality condition (3.14) follows from the facts that $\lambda + |\psi_y(t)| \neq 0$ $\forall t \in T$ and that matrix ξ'_y is nonsingular everywhere. The proof is complete. □

3.5 Second-Order Optimality Conditions for a Simple Problem

The idea underlying the method of investigation used in the previous section suggests that it is more convenient to derive the second-order necessary optimality conditions into two stages. While, in the first stage, the matrix G does not depend on x, in the second stage, the method of Sect. 3.4 is applied directly to derive these conditions in their full generality, albeit under the Frobenius condition. In order to implement this proposal, consider the following reduced impulsive control problem:

$$\text{Minimize } \varphi(p) \tag{3.30}$$

$$\text{subject to } dx = f(x,u,t)dt + G(t)d\mu, \tag{3.31}$$

$$e_1(p) \leq 0, \quad e_2(p) = 0, \tag{3.32}$$

$$\text{range}(\mu) \subset K, \ t \in T = [t_0, t_1],$$

where $p = (x(t_0), x(t_1))$, $x(t_0) = x_0$, $x(t_1) = x_1$, $e_i : \mathbb{R}^n \times \mathbb{R}^n \to \mathbb{R}^{d_i}$, $i = 1, 2$, are given twice continuously differentiable functions. As can be seen, in comparison with the formulation of Problem (3.1), the endpoint constraints $p \in S$ now take the reduced form (3.32), and, moreover, $U = \mathbb{R}^m$. Such requirements are dictated by the necessity to consider the second-order optimality conditions.[3]

Let us state the optimality necessary conditions for Problem (3.30)–(3.32). Denote:

$$L(p,\lambda) = \lambda^0\varphi(p) + \langle\lambda^1, e_1(p)\rangle + \langle\lambda^2, e_2(p)\rangle,$$

where $\lambda = (\lambda^0, \lambda^1, \lambda^2)$, $\lambda^0 \in \mathbb{R}^1$, $\lambda^i \in \mathbb{R}^{d_i}$, $i = 1, 2$. Note that function H remains the same as previously, while Q takes the form

$$Q(\psi, t) = G^*(t)\psi.$$

[3] The constraints on control $u(\cdot)$ can also be considered in the framework of investigation of the second-order optimality conditions; see, for example, [3, 9, 10, 20]. The assumption $U = \mathbb{R}^m$ is now considered for the sake of simplicity, as the main focus here is the impulsive control.

As was established earlier, the optimal process $(\hat{x}, \hat{u}, \hat{\mu})$ satisfies the Euler–Lagrange conditions[4]; that is, there exist a vector $\lambda \neq 0$, and an absolutely continuous vector-valued function ψ such that

$$\lambda^0 \geq 0, \quad \lambda^1 \geq 0, \quad \langle \lambda^1, e_1(\hat{p}) \rangle = 0, \tag{3.33}$$

$$\dot{\psi}(t) = -\frac{\partial H}{\partial x}(t), \quad \text{a.a. } t, \tag{3.34}$$

$$(\psi(t_0), -\psi(t_1)) = \frac{\partial L}{\partial p}(\hat{p}, \lambda), \tag{3.35}$$

$$\frac{\partial H}{\partial u}(t) = 0, \quad \text{a.a. } t, \tag{3.36}$$

$$\langle Q(t), v \rangle \leq 0 \ \forall v \in K, \quad \forall t \in T, \tag{3.37}$$

$$\int_T \left\langle Q(t), d\hat{\mu} \right\rangle = 0. \tag{3.38}$$

Denote by $\Lambda := \Lambda(\hat{x}, \hat{u}, \hat{\mu})$ the set of vectors λ which satisfy the Euler–Lagrange conditions (3.33)–(3.38) for the given process $(\hat{x}, \hat{u}, \hat{\mu})$. For convenience, all components of the inequality-type endpoint constraint $e_1(p) \leq 0$ corresponding to the nonactive indices (i.e., such indices j, that $e_1^j(\hat{p}) < 0$) are excluded, and in order to simplify the notation, the resulting vector-valued function is still denoted by e_1. In other words, in what follows, it is assumed that $e_1(\hat{p}) = 0$.

Introduce the notation

$$\mathscr{K}(\nu) = \text{lin}\{\nu\} + \left\{ \mu \in C^*(T; \mathbb{R}^k) : \text{range}(\mu) \subset K \right\},$$

where $\nu \in C^*(T; \mathbb{R}^k)$, and lin stands for the linear hull of a set. So, in particular, the set $\mathscr{K} := \mathscr{K}(0)$ is the cone of feasible control measures in Problem (3.30), i.e., $\mu \in \mathscr{K}$.

Consider the following differential equation with measure, which is usually termed *variational equation* or *variational system*:

$$d\delta x = \frac{\partial f}{\partial x}(t)\delta x dt + \frac{\partial f}{\partial u}(t)\delta u dt + G(t) d\delta\mu. \tag{3.39}$$

For each $\lambda \in \Lambda$, define the quadratic form

$$\Omega_\lambda : \mathbb{R}^n \times L_\infty(T; \mathbb{R}^m) \times C^*(T; \mathbb{R}^k) \to \mathbb{R},$$

[4]These conditions represent an essential part of the maximum principle. Note that, now, λ is not a number, but rather a vector, due to the new form of the endpoint constraints. Their set inclusion form (geometrical form) has been replaced by the classical inequality and equality type of constraints.

by the formula

$$\Omega_\lambda(\xi, \delta u, \delta\mu) = -\int_{t_0}^{t_1} \frac{\partial^2 H}{\partial(x,u)^2}(t)[(\delta x, \delta u)]^2 dt + \frac{\partial^2 L}{\partial p^2}(\hat{p}, \lambda)[\delta p]^2. \tag{3.40}$$

Here and below, δx designates the solution to (3.39) which corresponds to δu and $\delta\mu$ with initial condition $\delta x(t_0) = \xi$, while $Q[a]^2$ denotes the quadratic form $\langle a, Qa\rangle$.

Denote by Z the set of the triples $(\xi, \delta u, \delta\mu)$, such that $\delta\mu \in \mathscr{K}(\hat{\mu})$, and the endpoint vector $\delta p = (\delta x(t_0), \delta x(t_1))$ satisfies the boundary conditions

$$\frac{\partial e_1}{\partial p}(\hat{p})\delta p \le 0, \quad \frac{\partial e_2}{\partial p}(\hat{p})\delta p = 0, \tag{3.41}$$

$$\frac{\partial \varphi}{\partial p}(\hat{p})\delta p \le 0. \tag{3.42}$$

Set Z is, obviously, a cone. Its closure is known as the *cone of critical directions*.

Denote by N the linear subspace of all the triples $(\xi, \delta u, \delta\mu)$, such that

$$\delta\mu \in \mathscr{K} \cap (-\mathscr{K}) = \Big\{\mu \in C^*(T;\mathbb{R}^k) : \text{range}(\mu) \subset K \cap (-K)\Big\},$$

and Condition (3.41) is satisfied, in which, however, the inequality sign "≤" is replaced with equality "=". Note that $\delta\mu = 0$ over N when K is pointed.

Let the matrix Φ be the fundamental solution to (3.39); i.e., Φ is the solution to

$$\frac{d}{dt}\Phi(t) = \frac{\partial f}{\partial x}(t)\Phi(t), \quad \text{a.a. } t \in T, \quad \Phi(t_0) = I. \tag{3.43}$$

Consider the vector function $e = (e_1, e_2)$ and the *controllability matrix*

$$C := AA^* + \int_{t_0}^{t_1} B(t)B^*(t)dt,$$

where

$$A = \frac{\partial e}{\partial x_0}(\hat{p}) + \frac{\partial e}{\partial x_1}(\hat{p})\Phi(t_1), \quad B(t) = \frac{\partial e}{\partial x_1}(\hat{p})\Phi(t_1)\Phi^{-1}(t)\Big[\frac{\partial f}{\partial u}(t), G(t)P\Big].$$

Here, P stands for the orthogonal projector of $\mathbb{R}^k$ onto subspace $K \cap (-K)$. The matrix C is a square matrix of dimension $d \times d$, where $d = d_1 + d_2$. Let us set $q := \dim\ker C$.

Finally, consider the set of vectors $\lambda \in \Lambda$ such that the index[5] of Ω_λ over the subspace N is not greater than q, that is, $\text{ind}_N \Omega_\lambda \le q$. This set (a cone, generally

[5]Recall that the index of a quadratic form equals the maximal dimension of a subspace where the quadratic form is negative-definite.

nonconvex) is denoted by $\Lambda_a = \Lambda_a(\hat{x}, \hat{u}, \hat{\mu})$. It follows that Λ_a is nonempty at the weakest finite-dimensional minimizers. This property constitutes an essential part of the following second-order necessary conditions for optimality.

Theorem 3.3 *Let the admissible control process $(\hat{x}, \hat{u}, \hat{\mu})$ be a weakest (finite-dimensional) minimizer in Problem (3.30)–(3.32). Then, $\Lambda_a \neq \emptyset$, and for any $(\xi, \delta u, \delta\mu) \in Z$, it holds that*

$$\max_{\lambda \in \Lambda_a, |\lambda|=1} \Omega_\lambda(\xi, \delta u, \delta\mu) \geq 0. \tag{3.44}$$

The proof of this theorem is based on the application of the extremal principle obtained in [5]. Let us formulate it.

EXTREMUM PRINCIPLE. Let X be a vector space. Let $\mathscr{K}$ be a given convex cone in X. As usual, the normal cone to $\mathscr{K}$ at the point x is denoted by $N_{\mathscr{K}}(x)$. It stands for the set of all $x^* \in X^*$ such that $\langle x^*, y - x\rangle \leq 0 \; \forall\, y \in \mathscr{K}$, where X^* is the algebraically dual vector space, while the brackets signify the linear functional action. Let $F_i : X \to \mathbb{R}^{d_i}$, $i = 1, 2$ be given mappings and $\phi : X \to \mathbb{R}$ be a given functional.

Consider the minimization problem

$$\text{Minimize } \phi(x) \text{ subject to: } F_1(x) \leq 0, \; F_2(x) = 0, \; x \in \mathscr{K}. \tag{3.45}$$

Firstly, the notion of *finite topology* given on vector space X, [17], is introduced. In this topology, a set is open if its intersection with any finite-dimensional subspace M is open (in the unique Euclidean topology of M).[6] The finite topology on X is denoted by τ. We require that the cone $\mathscr{K}$ is closed w.r.t. τ.

Take a point $x_* \in X$. Without loss of generality, consider that $F_1(x_*) = 0$. Assume that the mappings ϕ and $F = (F_1, F_2) : X \to \mathbb{R}^d$, where $d = d_1 + d_2$, are twice continuously differentiable in a τ-neighborhood of point x_* w.r.t. τ. This means that, for an arbitrary finite-dimensional linear subspace $M \subseteq X$ such that $x_* \in M$, the restrictions of ϕ and F to M are twice continuously differentiable in a certain neighborhood O of x_* which depends on M. (For details, see [5, 11].)

Let us consider the Lagrange function

$$\mathscr{L}(x, \lambda) = \lambda^0 \phi(x) + \langle \lambda_1, F_1(x)\rangle + \langle \lambda_2, F_2(x)\rangle,$$

where $\lambda = (\lambda^0, \lambda_1, \lambda_2)$, $\lambda^0 \in \mathbb{R}$, $\lambda_1 \in \mathbb{R}^{d_1}$, and $\lambda_2 \in \mathbb{R}^{d_2}$. Let us denote by $\Lambda = \Lambda(x_*)$ the set of Lagrange multipliers λ: $\lambda^0 \geq 0$, $\lambda_1 \geq 0$, and $\lambda \neq 0$, which are associated with the point x_* by virtue of the Lagrange multiplier rule

---

[6] If space X is infinite-dimensional, then, after endowing it with the finite topology, it does not become a topological vector space as, in general, the addition is discontinuous. On the other hand, the finite topology is stronger than any other topology which can transform X into a vector topological space. Therefore, the minimum w.r.t. the finite topology is the weakest possible among all types of minima considered in optimal control.

$$\frac{\partial \mathscr{L}}{\partial x}(x_*, \lambda) \in -N_{\mathscr{K}}(x_*). \tag{3.46}$$

Given an arbitrary positive integer r, we consider the set of those Lagrange multipliers $\lambda \in \Lambda$ for which there exists a subspace $\Pi = \Pi(\lambda) \subseteq X$, such that

$$\operatorname{codim} \Pi \leq r, \quad \Pi \subseteq \ker F'(x_*), \quad \frac{\partial^2 \mathscr{L}}{\partial x^2}(x_*, \lambda)[x]^2 \geq 0 \;\; \forall x \in \Pi. \tag{3.47}$$

Above, codim stands for the codimension w.r.t. the subspace $\mathscr{K} \cap (-\mathscr{K})$, the maximal subspace contained in $\mathscr{K}$. This set of Lagrange multipliers is denoted by $\Lambda_r = \Lambda_r(x_*)$. Obviously, $\Lambda_r \subseteq \Lambda$, and Λ_r is also a cone.

Furthermore, consider the cone

$$Z(x_*) := \Big\{ h \in \operatorname{lin}\{x_*\} + \mathscr{K} : \big\langle \phi'(x_*), h \big\rangle \leq 0, \; F_1'(x_*)h \leq 0, \; F_2'(x_*)h = 0 \Big\},$$

whose closure is the cone of critical directions. Clearly, the cone $Z(x_*)$ is convex and, moreover, not empty.

Theorem 3.4 (Extremal principle) *Let point x_* be a local minimum w.r.t. the finite topology τ in Problem (3.45). Then, the set Λ_d is not empty and, moreover,*

$$\max_{\lambda \in \Lambda_d, |\lambda|=1} \frac{\partial^2 \mathscr{L}}{\partial x^2}(x_*, \lambda)[h]^2 \geq 0 \;\; \forall h \in Z(x_*).$$

This theorem represents a generalization of the second-order necessary conditions for abnormal problems obtained in Chap. 1 of [11]; see also in [4, 6] to the case of the extra infinite-dimensional cone constraint $x \in \mathscr{K}$. Its proof can be found in [5].

The proof of Theorem 3.3. The approach for the proof of Theorem 3.3 consists in decoding the conditions of the above-stated extremal principle in terms of the data of Problem (3.30)–(3.32). Firstly, formulate Problem (3.30)–(3.32) as an abstract extremal problem.

Note that, for any

$$(x_0, u, \mu) \in \mathscr{X} := \mathbb{R}^n \times \mathrm{L}_\infty(T; \mathbb{R}^m) \times \mathrm{C}^*(T; \mathbb{R}^k),$$

in some neighborhood of $(\hat{x}(t_0), \hat{u}, \hat{\mu})$, there exists the unique trajectory $x(\cdot)$ satisfying on T Eq. (3.31) with $x(t_0) = x_0$. It is not restrictive to assume that the minimizing functional φ depends only on x_0. Indeed, this can be inferred by a simple reduction.

Define

$$F_1(x_0, u, \mu) := e_1(x_0, x(t_1)), \quad F_2(x_0, u, \mu) := e_2(x_0, x(t_1)),$$

and consider the extremal problem

$$\text{Minimize } \; \varphi(x_0) \; \text{ subject to: } \; F_1(x_0, u, \mu) \leq 0, \; F_2(x_0, u, \mu) = 0,$$

$$x_0 \in \mathbb{R}^n, \ u \in L_\infty(T; \mathbb{R}^m), \ \mu \in \mathscr{K}.$$

The point $(\hat{x}(t_0), \hat{u}, \hat{\mu})$ is a minimum to this problem w.r.t. topology τ in view of Definition 3.3. Furthermore, it is a straightforward task to verify that the mappings F_1 and F_2 satisfy all conditions required for the application of the extremal principle.

Consider the Lagrange function

$$\mathscr{L}(x_0, u, \mu) = \lambda^0 \varphi(x_0) + \langle \lambda^1, F_1(x_0, u, \mu)\rangle + \langle \lambda^2, F_2(x_0, u, \mu)\rangle.$$

Next, apply the extremal principle. By calculating the derivatives $\frac{\partial \mathscr{L}}{\partial x_0}$, $\frac{\partial \mathscr{L}}{\partial u}$ and $\frac{\partial \mathscr{L}}{\partial \mu}$,[7] in view of (3.46), after some conventional transformation, see, e.g., [10], obtain

$$\frac{\partial L}{\partial x_0}(\hat{p}, \lambda) + \frac{\partial L}{\partial x_1}(\hat{p}, \lambda)\Phi(t_1) = 0, \tag{3.48}$$

$$\frac{\partial L}{\partial x_0}(\hat{p}, \lambda) \int_{t_0}^{t_1} \Phi^{-1}(t)\frac{\partial f}{\partial u}(t)\delta u(t)dt = 0 \ \ \forall\, \delta u \in L_\infty(T; \mathbb{R}^m), \tag{3.49}$$

$$\frac{\partial L}{\partial x_0}(\hat{p}, \lambda) \int_T \Phi^{-1}(t)G(t)d(\mu - \hat{\mu}) \le 0 \ \ \forall\, \mu \in \mathscr{K}. \tag{3.50}$$

By setting $\psi(t) := \frac{\partial L}{\partial x_0}(\hat{p}, \lambda)\Phi^{-1}(t)$, it can be concluded that the transversality condition in (3.35) follows from (3.48), Condition (3.36) follows from (3.49), and the maximum condition for the impulsive part, that is, Conditions (3.37), and (3.38), follows from (3.50). Moreover, it is easy to ensure that ψ satisfies the conjugate equation (3.34).

By restricting the arguments on the subspace N, and again, by following the same approach as in [10] (see the proof of Theorem 1 therein), it is easy to find that

$$\frac{\partial^2 \mathscr{L}(\hat{x}_0, \hat{u}, \hat{\mu})}{\partial (x_0, u, \mu)^2}[(\xi, \delta u, \delta\mu)]^2 = \Omega_\lambda(\xi, \delta u, \delta\mu) \ \ \forall\, (\xi, \delta u, \delta\mu) \in N,$$

where $\delta x(\cdot)$ is the solution to (3.39) w.r.t. $\delta x(t_0) = \xi$, δu, $\delta\mu$. The approach consists in considering the derivatives of Lagrange function along the direction $(\xi, \delta u, \delta\mu) \in N$. Then, the above equality on $(\xi, \delta u, \delta\mu)$ is proved by the Taylor series expansion upto the second order and by using the already derived first-order conditions (3.48)–(3.50).

By applying the extremal principle, from (3.47), find that there exists a subspace Π, such that

$$\operatorname{codim} \Pi \le d, \quad \Pi \subseteq N, \quad \Omega_\lambda(\xi, \delta u, \delta\mu) \ge 0 \ \ \forall\, (\xi, \delta u, \delta\mu) \in \Pi.$$

[7] They are computed in a standard way by using theorems on differentiation of the solution w.r.t. the parameter and initial data; see [16].

The codimension of $N = \ker F'(\hat{x}_0, \hat{u}, \hat{\mu}) \cap Y$, where

$$Y = \{(\xi, \delta u, \delta\mu) \in X : \delta\mu \in \mathscr{K} \cap (-\mathscr{K})\},$$

w.r.t. Y, equals $d - q$. (This is a standard task to verify, see, e.g., Proposition 1 in [9]. Observe that, along this line of arguments, it is not restrictive to consider measures merely absolutely continuous; see Exercise 3.9.) Then, the index of Ω_λ over N is not greater than q, and, therefore, $\Lambda_a \neq \varnothing$.

Finally, Condition (3.44) is the assertion of the extremal principle—Theorem 3.4. It is clear that the cone of critical directions $Z(x_*)$ from the extremal principle yields the cone Z of Theorem 3.3. Therefore, it has been verified that Theorem 3.3 results from the application of the extremal principle to the impulsive control problem (3.30).

The proof is complete. □

3.6 Second-Order Necessary Conditions Under the Frobenius Condition

By combining the two approaches—from Sect. 3.4 and from Sect. 3.5—the second-order necessary optimality conditions for the case, in which the matrix-valued function G depends on x, are derived under the Frobenius condition as below.

Consider the following problem

$$\begin{aligned} &\text{Minimize } \ \varphi(p) \\ &\text{subject to: } \ dx = f(x,u,t)dt + G(x,t)d\mu, \\ &\qquad e_1(p) \le 0, \ \ e_2(p) = 0, \\ &\qquad \text{range}(\mu) \subset K, \ \ t \in T = [t_0, t_1]. \end{aligned} \tag{3.51}$$

The only difference between this problem and (3.30) is that Eq. (3.31) is now more general: The matrix G depends on x.

Let us formulate the necessary conditions of optimality for Problem (3.51). Consider the optimal process $(\hat{x}, \hat{u}, \hat{\mu})$ in Problem (3.51). As is already known, it satisfies the Euler–Lagrange conditions. Then, there exists a vector $\lambda = (\lambda^0, \lambda^1, \lambda^2) \neq 0$, and a vector-valued function of bounded variation $\psi(\cdot)$ such that Conditions (3.33), (3.35)–(3.38), and the conjugate equation in (3.15) are satisfied. As before, denote by $\Lambda := \Lambda(\hat{x}, \hat{u}, \hat{\mu})$ the set of all vectors λ that satisfy the Euler–Lagrange conditions. (Note that ψ can be uniquely determined by $\lambda \in \Lambda$.) As before, assume also that $e_1(\hat{p}) = 0$.

Consider the variational differential equation with measure:

$$d\delta x = \frac{\partial f}{\partial x}(t)\delta x dt + \frac{\partial f}{\partial u}(t)\delta u dt + G(t)d\delta\mu + \frac{\partial G}{\partial x}(t)\delta x d\hat{\mu}. \tag{3.52}$$

It is necessary to clarify the meaning of solution to (3.52). Note that this equation contains two vector measures: $\hat{\mu}$ and $\delta\mu$. However, the solution is defined in the same way as the trajectory. The vector-valued function of bounded variation $\delta x(t)$ is said to be the solution to (3.52) provided that, for all $t > t_0$,

$$\delta x(t) = \delta x(t_0) + \int_{t_0}^{t} \frac{\partial f}{\partial x}(\varsigma)\delta x(\varsigma)d\varsigma + \int_{t_0}^{t} \frac{\partial f}{\partial u}(\varsigma)\delta u(\varsigma)d\varsigma + \int_{[t_0,t]} G(\varsigma)d\delta\mu_{\mathrm{c}}$$
$$+ \int_{[t_0,t]} \frac{\partial G}{\partial x}(\varsigma)\delta x(\varsigma)d\hat{\mu}_{\mathrm{c}} + \sum_{\tau\in\mathrm{Ds}((\hat{\mu},\delta\mu)):\,\tau\le t} \Big(\delta x_\tau(1) - \delta x(\tau^-)\Big),$$

where $\delta x_\tau(\cdot)$ satisfies on $[0, 1]$ the attached variational differential equation

$$\dot{\delta x}_\tau = G(\hat{x}_\tau(s), \tau)\delta\mu(\tau) + G'_x(\hat{x}_\tau(s), \tau)\delta x_\tau \hat{\mu}(\tau).$$

For each $\lambda \in \Lambda$, define the quadratic form $\Omega_\lambda : \mathbb{R}^n \times \mathrm{L}_\infty(T; \mathbb{R}^m) \times C^*(T; \mathbb{R}^k) \to \mathbb{R}$, by the formula

$$\Omega_\lambda(\xi, \delta u, \delta\mu) = -\int_{t_0}^{t_1} \frac{\partial^2 H}{\partial(x,u)^2}(t)[(\delta x, \delta u)]^2 dt - 2\int_T \Big\langle \frac{\partial Q}{\partial x}(t)\delta x, d\delta\mu \Big\rangle -$$
$$\int_T \Big\langle \frac{\partial^2 Q}{\partial x^2}(t)[\delta x]^2, d\hat{\mu} \Big\rangle + \frac{\partial^2 L}{\partial p^2}(\hat{p}, \lambda)[\delta p]^2. \tag{3.53}$$

Here and in that which follows, δx is the solution to (3.52) that corresponds to δu and $\delta\mu$ with initial condition $\delta x(t_0) = \xi$, and $Q[a]^2$ denotes the quadratic form $\langle a, Qa\rangle$. The meaning of the second and third integrals in (3.53) has also to be clarified as they are not classical Lebesgue integrals. However, the following convention is adopted:

$$\int_T \Big\langle \frac{\partial Q}{\partial x}(t)\delta x, d\delta\mu \Big\rangle =$$
$$= \int_T \Big\langle \frac{\partial Q}{\partial x}(t)\delta x, d\delta\mu_{\mathrm{c}} \Big\rangle + \sum_{\tau\in\mathrm{Ds}(\delta\mu)} \int_0^1 \Big\langle \frac{\partial Q}{\partial x}(\hat{x}_\tau(s), \psi_\tau(s), \tau)\delta x_\tau(s), \delta\mu(\tau) \Big\rangle ds,$$

$$\int_T \Big\langle \frac{\partial^2 Q}{\partial x^2}(t)[\delta x]^2, d\hat{\mu} \Big\rangle =$$
$$= \int_T \Big\langle \frac{\partial^2 Q}{\partial x^2}(t)[\delta x]^2, d\hat{\mu}_{\mathrm{c}} \Big\rangle + \sum_{\tau\in\mathrm{Ds}(\hat{\mu})} \int_0^1 \Big\langle \frac{\partial^2 Q}{\partial x^2}(\hat{x}_\tau(s), \psi_\tau(s), \tau)[\delta x_\tau(s)]^2, \hat{\mu}(\tau) \Big\rangle ds.$$

Denote by Z the cone of critical directions. That is, as previously, the set of the triples $(\xi, \delta u, \delta\mu)$, such that $\delta\mu \in \mathscr{K}(\hat{\mu})$, and the endpoint vector $\delta p = (\delta x(t_0), \delta x(t_1))$ satisfy the boundary conditions in (3.41), (3.42). Denote by N, as previously the linear subspace of all the triples $(\xi, \delta u, \delta\mu) \in Y$, such that $\frac{\partial e}{\partial p}(\hat{p})\delta p = 0$.

Let matrix Φ be the fundamental solution to (3.52). This means that Φ is the solution to the following differential equation with measure

$$d\Phi(t) = \frac{\partial f}{\partial x}(t)\Phi(t)dt + \frac{\partial G}{\partial x}(t)\Phi(t)d\hat{\mu}, \quad \Phi(t_0) = I. \tag{3.54}$$

(Compare it with the conventional equation (3.43) which does not involve a measure.) This solution is not conventional, and, therefore, its definition requires clarification as does the solution to (3.52). Matrix-valued function of bounded variation $\Phi(t)$ is said to be the solution to (3.54), provided that $\Phi(t_0) = I$, and, for all $t > t_0$,

$$\Phi(t) = I + \int_{t_0}^{t} \frac{\partial f}{\partial x}(\varsigma)\Phi(\varsigma)d\varsigma + \int_{[t_0,t]} \frac{\partial G}{\partial x}(\varsigma)\Phi(\varsigma)d\hat{\mu}_c + \sum_{\tau\in \mathrm{Ds}(\hat{\mu}):\tau\le t} \Big[\Phi_\tau(1) - \Phi(\tau^-)\Big],$$

where matrix function $\Phi_\tau(\cdot)$ satisfies on $[0, 1]$ the equation

$$\dot{\Phi}_\tau = G'_x(\hat{x}_\tau(s), \tau)\Phi_\tau\hat{\mu}(\tau).$$

Let us consider the matrices A and B and the associated controllability matrix C as defined in Sect. 3.5, but now w.r.t. $G = G(x, t)$. The matrix C is a square matrix of dimension $d \times d$. Let us set $q = \dim\ker C$. The set of all vectors $\lambda \in \Lambda$ such that $\mathrm{ind}_N\,\Omega_\lambda \le q$ is denoted by $\Lambda_a = \Lambda_a(\hat{x}, \hat{u}, \hat{\mu})$.

Theorem 3.5 *Let the admissible control process $(\hat{x}, \hat{u}, \hat{\mu})$ be a weakest (finite-dimensional) minimizer in Problem (3.51). Then, $\Lambda_a \ne \varnothing$, and for any $(\xi, \delta u, \delta\mu) \in Z$, it holds that*

$$\max_{\lambda\in\Lambda_a, |\lambda|=1} \Omega_\lambda(\xi, \delta u, \delta\mu) \ge 0. \tag{3.55}$$

Proof The proof of this theorem consists in applying the reduction presented in Sect. 3.4, and, subsequently, in applying the result of Sect. 3.5, that is, Theorem 3.3, to the reduced problem.

Let $G = G(x)$. Along with the original problem (3.51), consider the following reduced impulsive control problem.

$$\begin{aligned}
\text{Minimize } \ & \varphi(y_0, \xi(w_1, y_1)) \\
\text{subject to: } \ & dy = g(w, y, u, t)dt, \\
& dw = d\mu, \\
& w_0 = 0,\ e_1(y_0, \xi(w_1, y_1)) \le 0,\ e_2(y_0, \xi(w_1, y_1)) = 0, \\
& \mathrm{range}(\mu) \subset K,\ t \in T.
\end{aligned} \tag{3.56}$$

Here,

$$g(w, y, u, t) := \eta'_x(w, \xi(w, y)) f(\xi(w, y), u, t),$$

and $y_0 = y(t_0)$, $y_1 = y(t_1)$, $w_0 = w(t_0)$, and $w_1 = w(t_1)$.

Problems (3.51) and (3.56) are equivalent. This assertion readily follows from Lemma 3.1 as, in view of (3.10), the admissible processes of the considered problems are related by the following one-to-one correspondence:

$$(x, u, \mu) \rightarrowtail (y = \eta(w, x), w, u, \mu), \quad \text{and} \quad (y, w, u, \mu) \rightarrowtail (x = \xi(w, y), u, \mu),$$

which, obviously, preserves the value of the cost as well as the endpoint constraints.

Let the process $(\hat{x}, \hat{u}, \hat{\mu})$ be optimal in (3.51). Then, the process $(\hat{y} = \eta(\hat{w}, \hat{x}), \hat{w}, \hat{u}, \hat{\mu})$, where $\hat{w}(t) = F(t; \hat{\mu})$, is optimal in (3.56). Let us apply to it Theorem 3.3. The functions H, Q, L take the following form

$$\begin{aligned} &\tilde{H}(y, w, u, \psi_y, \psi_w, t) := \langle \psi_y, g(w, y, u, t) \rangle, \\ &\tilde{Q}(\psi_w) := \psi_w, \\ &\tilde{L}(z, \lambda) := \lambda^0 \varphi(y_0, \xi(w_1, y_1)) + \langle \lambda^1, e_1(y_0, \xi(w_1, y_1)) \rangle + \langle \lambda^2, e_2(y_0, \xi(w_1, y_1)) \rangle, \end{aligned}$$

where $z = (y_0, y_1, w_1)$.

Then, there exist a vector $\lambda = (\lambda^0, \lambda^1, \lambda^2) \neq 0$, and absolutely continuous row vector functions ψ_y, ψ_w such that the Euler–Lagrange conditions are satisfied:

$$\begin{aligned} \dot{\psi}_y(t) &= -\tilde{H}'_y(t), \\ \dot{\psi}_w(t) &= -\tilde{H}'_w(t), \end{aligned} \tag{3.57}$$

$$(\psi_y(t_0), -\psi_y(t_1), -\psi_w(t_1)) = \tilde{L}'_z(\hat{z}, \lambda), \tag{3.58}$$

$$\tilde{H}'_u(t) = 0 \ \text{a.e.}, \tag{3.59}$$

$$\langle \psi_w(t), v \rangle \leq 0 \ \ \forall v \in K, \ \ \forall t \in T, \tag{3.60}$$

$$\int_T \langle \psi_w(t), d\hat{\mu} \rangle = 0. \tag{3.61}$$

Here, $\hat{z} = (\hat{y}_0, \hat{y}_1, \hat{w}_1)$.

Furthermore, the second-order conditions are satisfied as follows. Consider the variational system

$$\begin{aligned} \dot{\delta y} &= g'_y(t)\delta y + g'_w(t)\delta w + g'_u(t)\delta u, \\ d\delta w &= d\delta\mu. \end{aligned} \tag{3.62}$$

Over the solutions $\delta y, \delta w$ to (3.62) corresponding to $\delta u, \delta\mu$ and the initial value $\delta y_0 = \delta y(t_0) = \xi$, $\delta w(t_0) = 0$, consider the quadratic form

$$\tilde{\Omega}_\lambda(\xi, \delta u, \delta\mu) = -\int_{t_0}^{t_1} \frac{\partial^2 \tilde{H}(t)}{\partial(y, w, u)^2}[(\delta y, \delta w, \delta u)]^2 dt + \frac{\partial^2 \tilde{L}}{\partial z^2}(\hat{z}, \lambda)[\delta z]^2,$$

where $\delta z = (\delta y_0, \delta y_1, \delta w_1)$.

Denote by $\tilde{Z}$ the set of the triples $(\xi, \delta u, \delta\mu)$ such that $\delta\mu \in \mathscr{K}(\hat{\mu})$, and the endpoint vector δz satisfies the boundary conditions

$$\frac{\partial \tilde{e}_1}{\partial z}(\hat{z})\delta z \le 0, \quad \frac{\partial \tilde{e}_2}{\partial z}(\hat{z})\delta z = 0,$$

$$\frac{\partial \tilde{\varphi}}{\partial z}(\hat{z})\delta z \le 0.$$

Above, $z = (y_0, y_1, w_1)$, $\tilde{e}_1(z) = e_1(y_0, \xi(w_1, y_1))$, $\tilde{e}_2(z) = e_2(y_0, \xi(w_1, y_1))$, and $\tilde{\varphi}(z) = \varphi(y_0, \xi(w_1, y_1))$. The linear subspace of all the triples $(\xi, \delta u, \delta\mu)$, such that $\delta\mu \in \mathscr{K} \cap (-\mathscr{K})$ and $\frac{\partial \tilde{e}}{\partial z}(\hat{z})\delta z = 0$, where $\tilde{e} = (\tilde{e}_1, \tilde{e}_2)$ is denoted by $\tilde{N}$.

Let $\tilde{\Phi} : T \to \mathbb{R}^{(n+k)\times(n+k)}$ be the fundamental solution to (3.62), that is,

$$\frac{d}{dt}\tilde{\Phi}(t) = \begin{bmatrix} g'_y(t) & g'_w(t) \\ 0 & 0 \end{bmatrix}\tilde{\Phi}(t), \quad \tilde{\Phi}(t_0) = I.$$

Consider the controllability matrix

$$\tilde{C} := \tilde{A}\tilde{A}^* + \int_{t_0}^{t_1} \tilde{B}(t)\tilde{B}^*(t)dt,$$

where

$$\tilde{A} = \frac{\partial \tilde{e}(\hat{z})}{\partial y_0} + \frac{\partial \tilde{e}(\hat{z})}{\partial y_1}\tilde{\Phi}_y(t_1),$$

$$\tilde{B}(t) = \frac{\partial \tilde{e}(\hat{z})}{\partial(y_1, w_1)}\tilde{\Phi}(t_1)\tilde{\Phi}^{-1}(t)\begin{bmatrix} g'_u(t) & 0 \\ 0 & P \end{bmatrix}.$$

Here, $\tilde{\Phi}_y$ is the first $n \times n$ block of $\tilde{\Phi}$.

Let $\tilde{q} := \dim\ker \tilde{C}$. Consider the set of vectors $\lambda \in \tilde{\Lambda}$ such that $\operatorname{ind}_{\tilde{N}} \tilde{\Omega}_\lambda \le q$, where $\tilde{\Lambda}$ is the set of nontrivial Lagrange multipliers satisfying (3.57)–(3.61). This set is denoted by $\tilde{\Lambda}_a$. The second-order optimality condition states that $\tilde{\Lambda}_a \ne \varnothing$, and for any $(\xi, \delta u, \delta\mu) \in \tilde{Z}$, it holds that

$$\max_{\lambda\in\tilde{\Lambda}_a, |\lambda|=1} \tilde{\Omega}_\lambda(\xi, \delta u, \delta\mu) \ge 0. \tag{3.63}$$

After stating the optimality conditions for the reduced problem (3.56), let us deduce (3.55) from (3.63) by establishing an appropriate one-to-one correspondence between the Lagrange multipliers and, also, between the variational solutions to both problems. The correspondence between the conjugate functions is given by

$$\psi(t) := \psi_y(t)\eta'_x(\hat{w}(t), \hat{x}(t)) = \psi_y(t)[\xi'_y(\hat{w}(t), \hat{y}(t))]^{-1}.$$

Then, from (3.57)–(3.61), it follows that $\lambda \in \Lambda$. Indeed, this has already been shown in the proof of Theorem 3.2. The only small difference concerns the decoding of (3.58). Note that, after calculating the derivative of the composition of functions, the transversality condition (3.58) yields

$$(\psi_y(t_0), -\psi_y(t_1)) \in \Big(L'_{x_0}(\hat{p}, \lambda), L'_{x_1}(\hat{p}, \lambda)\xi'_y(\hat{w}_1, \hat{y}_1)\Big),$$

$$\psi_w(t_1) = -L'_{x_1}(\hat{p}, \lambda)\xi'_w(\hat{w}_1, \hat{y}_1)) = -L'_{x_1}(\hat{p}, \lambda)G(t_1).$$

Note that, since the correspondence is one-to-one it can actually be proven that $\tilde{\Lambda} = \Lambda$.

Ensure that $\tilde{\Lambda}_a = \Lambda_a$ and also that Condition (3.55) is satisfied. Firstly, let us establish a one-to-one correspondence between the variational solutions δx and δy.

Consider the variable change

$$\delta x(t) = \xi'_y(\hat{w}(t), \hat{y}(t))\delta y(t) + \xi'_w(\hat{w}(t), \hat{y}(t))\delta w(t) \tag{3.64}$$

and show that δx satisfies variational differential equation with measure (3.52). This will be performed firstly for the case when both measures $\delta\mu$ and $\hat{\mu}$ are absolutely continuous, i.e., when $\delta\mu = \delta m(t)dt$ and $\hat{\mu} = \hat{m}(t)dt$.

By differentiating, obtain

$$\begin{aligned}
&\dot{\delta x}(t) = \tfrac{d}{dt}\Big(\xi'_y(\hat{w}(t), \hat{y}(t))\Big)\delta y(t) + \xi'_y(\hat{w}(t), \hat{y}(t))\dot{\delta y}(t) + \tfrac{d}{dt}\Big(G(\hat{x}(t))\Big)\delta w(t) + \\
&G(\hat{x}(t))\delta m(t) = \Big(\xi''_{yw}(t)\hat{m}(t) + \xi''_{yy}(t)\eta'_x(t)f(t)\Big)\delta y(t) + \\
&\xi'_y(t)\Big(g'_y(t)\delta y(t) + g'_w(t)\delta w(t) + g'_u(t)\delta u(t)\Big) + G'_x(t)\Big(f(t) + G(t)\hat{m}(t)\Big)\delta w(t) + \\
&G(\hat{x}(t))\delta m(t) = G'_x(t)\xi'_y(t)\delta y(t)\hat{m}(t) - \xi'_y(t)\eta''_{xx}(t)\xi'_y(t)f(t)\delta y(t) + \\
&\xi'_y(t)\eta''_{xx}(t)\xi'_y(t)f(t)\delta y(t) + \xi'_y(t)\eta'_x(t)f'_x(t)\xi'_y(t)\delta y(t) + \\
&\xi'_y(t)(\eta''_{xw}(t) + \eta''_{xx}(t)G(t))f(t)\delta w(t) + \xi'_y(t)\eta'_x(t)f'_x(t)G(t)\delta w(t) + \xi'_y(t)\eta'_x(t)f'_u(t)\delta u(t) + \\
&G'_x(t)f(t)\delta w(t) + G'_x(t)G(t)\hat{m}(t)\delta w(t) + G(\hat{x}(t))\delta m(t) = G'_x(t)\xi'_y(t)\delta y(t)\hat{m}(t) + \\
&f'_x(t)\xi'_y(t)\delta y(t) - \xi'_y(t)\eta'_x(t)G'_x(t)f(t)\delta w(t) + f'_x(t)G(t)\delta w(t) + f'_u(t)\delta u(t) + \\
&G'_x(t)f(t)\delta w(t) + G'_x(t)G(t)\hat{m}(t)\delta w(t) + G(\hat{x}(t))\delta m(t) = G'_x(t)\delta x(t)\hat{m}(t) - \\
&G'_x(t)G(t)\delta w(t)\hat{m}(t) + f'_x(t)\delta x(t) - f'_x(t)G(t)\delta w(t) - G'_x(t)f(t)\delta w(t) + \\
&f'_x(t)G(t)\delta w(t) + f'_u(t)\delta u(t) + G'_x(t)f(t)\delta w(t) + G'_x(t)G(t)\hat{m}(t)\delta w(t) + \\
&G(\hat{x}(t))\delta m(t) = G'_x(t)\delta x(t)\hat{m}(t) + f'_x(t)\delta x(t) + f'_u(t)\delta u(t) + G(\hat{x}(t))\delta m(t).
\end{aligned}$$

Above, the following have been used successively: the fact that $\xi'_w = G(\xi)$, (3.62), Exercise 3.10, Conditions (3.8), (3.28), (3.64), and, in the last equality, the symmetry of the bilinear map $G'_x(t)G(t)$ due to the Frobenius condition. The dependence of the functions on x, y, w has been omitted after the first line, for convenience of reading.

So, the variational equation in (3.52) is valid for absolutely continuous measures. When $\hat{\mu}$, $\delta\mu$ are arbitrary measures, this equation still holds true as these measures can be weakly-* approximated by absolutely continuous measures. Note that, in view of the formula (3.64), the jumps of $\delta x(t)$ do not depend on the method by which the control measure is approximated by absolutely continuous measures. Thus, by appropriate choice of the approximating sequence—as in the proof of Proposition 3.1—Equation (3.52) can be derived in its full generality. See Exercise 3.11.

Consider the fundamental matrix $\tilde{\Phi}$. Note that it is a block matrix, such that

$$\tilde{\Phi} = \begin{bmatrix} \tilde{\Phi}_y & \tilde{\Phi}_w \\ 0 & I \end{bmatrix}.$$

Let us verify that $\Phi(t) = \xi'_y(\hat{w}(t), \hat{y}(t))\tilde{\Phi}_y(t)$. Indeed, by once more assuming that measure $\hat{\mu}$ is absolutely continuous, and by taking into account Exercise 3.10, we have

$$\begin{aligned}
&\dot{\Phi}(t) = \tfrac{d}{dt}(\xi'_y(\hat{w}(t), \hat{y}(t))\tilde{\Phi}_y(t) + \xi'_y(\hat{w}(t), \hat{y}(t))\dot{\tilde{\Phi}}_y(t) = \\
&\left(\xi''_{yw}(t)\hat{m}(t) + \xi''_{yy}(t)\eta'_x(t)f(t)\right)\tilde{\Phi}_y(t) + \xi'_y(t)g'_y(t)\tilde{\Phi}_y(t) = \\
&G'_x(t)\xi'_y(t)\tilde{\Phi}_y(t)\hat{m}(t) + \xi''_{yy}(t)\eta'_x(t)f(t)\tilde{\Phi}_y(t) + \xi'_y(t)\eta''_{xx}(t)\xi'_y(t)f(t)\tilde{\Phi}_y(t) + \\
&\xi'_y(t)\eta'_x(t)f'_x(t)\xi'_y(t)\tilde{\Phi}_y(t) = G'_x(t)\Phi(t)\hat{m}(t) + f'_x(t)\Phi(t).
\end{aligned}$$

Thus, (3.54) is proved. The general case is treated as above. Note that, again, the solution Φ will not depend on the method of approximation by absolutely continuous measures. Then, it clearly holds that $\tilde{A} = A$. It is also a straightforward task to ensure that $\tilde{B}(t) = B(t)$. Indeed, this is due to the following simple relation between the state transition matrices $S(t,\tau) = \Phi(t)\Phi^{-1}(\tau)$ and $\tilde{S}(t,\tau) = \tilde{\Phi}(t)\tilde{\Phi}^{-1}(\tau)$:

$$S(t,\tau) = \Xi(t)\tilde{S}(t,\tau)[\Xi^*(t)\Xi(t)]^{-1}\Xi^*(t), \quad \text{where } \Xi(t) = [\xi'_y(t), \xi'_w(t)].$$

This equality follows directly from (3.64) and the properties of state transition matrices. Note that $\Xi^*(t)\Xi(t)$ is invertible as $\xi'_y(t)$ is nonsingular. Then,

$$\begin{aligned}
&\tilde{B}(t) = \tfrac{\partial\tilde{e}(\hat{z})}{\partial(y_1,w_1)}\tilde{S}(t_1,t)\begin{bmatrix} g'_u(t) & 0 \\ 0 & P \end{bmatrix} = \\
&\tfrac{\partial e}{\partial x_1}(\hat{p})\Xi(t_1)\tilde{S}(t_1,t)[\Xi^*(t)\Xi(t)]^{-1}\Xi^*(t)\Xi(t)\begin{bmatrix} g'_u(t) & 0 \\ 0 & P \end{bmatrix} = \tfrac{\partial e}{\partial x_1}(\hat{p})S(t_1,t)[f'_u(t), G(t)P],
\end{aligned}$$

and, thus, $\tilde{C} = C$, $\tilde{q} = q$. It is also clear that, in view of (3.64), $\tilde{Z} = Z$, $\tilde{N} = N$.

Now, the most difficult task remains: to show that $\Omega_\lambda \equiv \tilde{\Omega}_\lambda$.

Firstly, let us compute the second form of the Lagrangian L.

$$\begin{aligned}
&\tilde{L}''_{zz}(\hat{z},\lambda)[\delta z]^2 = L''_{x_0x_0}(\hat{p},\lambda)[\delta y_0]^2 + L''_{x_0x_1}(\hat{p},\lambda)[\delta y_0, \xi'_y(t_1)\delta y_1]+\\
&L''_{x_0x_1}(\hat{p},\lambda)[\delta y_0, G(t_1)\delta w_1] + L''_{x_1x_0}(\hat{p},\lambda)[\xi'_y(t_1)\delta y_1, \delta y_0] + L''_{x_1x_1}(\hat{p},\lambda)[\xi'_y(t_1)\delta y_1]^2+\\
&L'_{x_1}(\hat{p},\lambda)\xi''_{yy}[\delta y_1]^2 + L''_{x_1x_1}(\hat{p},\lambda)[\xi'_y(t_1)\delta y_1, G(t_1)\delta w_1] + L'_{x_1}(\hat{p},\lambda)\xi''_{yw}[\delta y_1, \delta w_1]+\\
&L''_{x_1x_0}(\hat{p},\lambda)[G(t_1)\delta w_1, \delta y_0] + L''_{x_1x_1}(\hat{p},\lambda)[G(t_1)\delta w_1, \xi'_y(t_1)\delta y_1] + L'_{x_1}(\hat{p},\lambda)\xi''_{wy}[\delta w_1, \delta y_1]+\\
&L''_{x_1x_1}(\hat{p},\lambda)[G(t_1)\delta w_1]^2 + L'_{x_1}(\hat{p},\lambda)\xi''_{ww}[\delta w_1]^2 = L''_{x_0x_0}(\hat{p},\lambda)[\delta x_0]^2+\\
&2L''_{x_0x_1}(\hat{p},\lambda)[\delta y_0, \delta x_1] + L''_{x_1x_1}(\hat{p},\lambda)[\delta x_1]^2 + L'_{x_1}(\hat{p},\lambda)\xi''(t_1)[(\delta y_1, \delta w_1)]^2 =\\
&L''_{pp}(\hat{p},\lambda)[\delta p]^2 + L'_{x_1}(\hat{p},\lambda)\xi''(t_1)[(\delta y_1, \delta w_1)]^2.
\end{aligned}$$

Let us calculate the second-order derivative of g. (Below, the dependence on arguments is omitted in order to simplify the notation.)

$$\begin{aligned}
g''_{yy} &= (\eta''_{xx}\xi'_y f + \eta'_x f'_x\xi'_y)'_y = (\eta'_x f'_x\xi'_y - \eta'_x\xi''_{yy}\eta'_x f)'_y = \eta'_x f''_{xx}\xi'_y\xi'_y + \eta'_x f'_x\xi''_{yy}+\\
&\quad 2\eta'_x\xi''_{yy}\eta'_x\xi''_{yy}\eta'_x f - \eta'_x\xi'''_{yyy}\eta'_x f - 2\eta'_x\xi''_{yy}\eta'_x f'_x\xi'_y;\\
g''_{yw} &= (\eta''_{xx}\xi'_y f + \eta'_x f'_x\xi'_y)'_w = (\eta'_x f'_x\xi'_y - \eta'_x\xi''_{yy}\eta'_x f)'_w = \eta'_x f''_{xx}G\xi'_y + \eta'_x f'_x\xi''_{yw} -\\
&\quad \eta'_x G'_x f'_x\xi'_y + \eta'_x G'_x\xi''_{yy}\eta'_x f - \eta'_x\xi'''_{yyw}\eta'_x f + \eta'_x\xi''_{yy}\eta'_x G'_x f - \eta'_x\xi''_{yy}\eta'_x f'_x G;\\
g''_{ww} &= ((\eta''_{xw} + \eta''_{xx}\xi'_w)f + \eta'_x f'_x\xi'_w)'_w = (\eta'_x f'_x\xi'_w - \eta'_x G'_x f)'_w = \eta'_x f''_{xx}GG + \eta'_x f'_x\xi''_{ww} -\\
&\quad \eta'_x G'_x f'_x G + \eta'_x G'_x G'_x f - \eta'_x G''_{xx}Gf - \eta'_x G'_x f'_x G = \eta'_x f''_{xx}GG + \eta'_x f'_x\xi''_{ww} -\\
&\quad 2\eta'_x G'_x f'_x G + 2\eta'_x G'_x G'_x f - \eta'_x\xi'''_{yww}\eta'_x f;\\
g''_{uu} &= \eta'_x f''_{uu};\\
g''_{uy} &= (\eta'_x f'_u)'_y = \eta''_{xx}\xi'_y f'_u + \eta'_x f''_{ux}\xi'_y = \eta'_x f''_{ux}\xi'_y - \eta'_x\xi''_{yy}\eta'_x f'_u;\\
g''_{uw} &= (\eta'_x f'_u)'_w = \eta'_x f''_{ux}G - \eta'_x G'_x f'_u.
\end{aligned}$$

Above, (3.28), Exercise 3.10, and the relation $\xi'''_{yww}\eta'_x = G'_x G'_x + G''_{xx}G$ have been used. Then, the second-order derivative equals

$$\begin{aligned}
g''(\delta y, \delta w, \delta u) = g''_{yy}[\delta y]^2 + g''_{ww}[\delta w]^2 + g''_{uu}[\delta u]^2 +\\
2g''_{yw}[\delta y, \delta w] + 2g''_{uy}[\delta u, \delta y] + 2g''_{uw}[\delta u, \delta w].
\end{aligned}$$

Let us gather terms in various "semantic" groups. Let us denote

$$\begin{aligned}
A_1(\delta y, \delta w, \delta u) &= f''_{xx}[\xi'_y\delta y]^2 + 2f''_{xx}[\xi'_y\delta y, G\delta w] + f''_{xx}[G\delta w]^2 +\\
&\quad 2f''_{xu}[\xi'_y\delta y, \delta u] + 2f''_{xu}[G\delta w, \delta u] + f''_{uu}[\delta u]^2;\\
A_2(\delta y, \delta w) &= f'_x(\xi''_{yy}[\delta y]^2 + 2\xi''_{yw}[\delta y, \delta w] + 2\xi''_{ww}[\delta w]^2) = f'_x\xi''(\delta y, \delta w)\\
A_3(\delta y, \delta w) &= -\xi'''_{yyy}\eta'_x f[\delta y]^2 - 2\xi'''_{yyw}\eta'_x f[\delta y, \delta w] - \xi'''_{yww}\eta'_x f[\delta w]^2;\\
A_4(\delta y, \delta w) &= 2\xi''_{yy}\eta'_x\xi''_{yy}\eta'_x f[\delta y]^2 + 2\xi''_{yy}\eta'_x\xi''_{yw}\eta'_x f[\delta y, \delta w]+\\
&\quad 2\xi''_{yw}\eta'_x\xi''_{yy}\eta'_x f[\delta w, \delta y] + 2\xi''_{yw}\eta'_x\xi''_{yw}\eta'_x f[\delta w]^2;\\
A_5(\delta y, \delta w, \delta u) &= -2\xi''_{yy}\eta'_x f'_x[\delta y, \xi'_y\delta y] - 2\xi''_{yw}\eta'_x f'_x[\delta w, \xi'_y\delta y] -\\
&\quad 2\xi''_{yy}\eta'_x f'_x[\delta y, G\delta w] - 2\xi''_{yw}\eta'_x f'_x[\delta w, G\delta w] -\\
&\quad 2\xi''_{yy}\eta'_x f'_u[\delta y, \delta u] - 2\xi''_{yw}\eta'_x f'_u[\delta w, \delta u].
\end{aligned}$$

In view of relation $G'_x = \xi''_{yw}\eta'_x$, it is clear that

$$g'' = \eta'_x\Big(A_1 + A_2 + A_3 + A_4 + A_5\Big). \tag{3.65}$$

In view of properties of ξ, for $\delta w_0 = 0$, we have $L'_{x_0}(\hat{p}, \lambda)\xi''(t_0)[(\delta y_0, \delta w_0)]^2 = 0$. Then, by taking into account the transversality condition (3.35),

$$L'_{x_1}(\hat{p}, \lambda)\xi''(t_1)[(\delta y_1, \delta w_1)]^2 + L'_{x_0}(\hat{p}, \lambda)\xi''(t_0)[(\delta y_0, \delta w_0)]^2 =$$
$$-\Big(\psi(t_1)\xi''(t_1)[(\delta y_1, \delta w_1)]^2 - \psi(t_0)\xi''(t_0)[(\delta y_0, \delta w_0)]^2\Big) =$$
$$-\int_{t_0}^{t_1} d\Big(\psi(t)\xi''(t)[(\delta y(t), \delta w(t))]^2\Big). \tag{3.66}$$

The next step is to apply the formula of integration by parts, though not the classical one, as in the Riemann–Stieltjes sense, but the one which is robust w.r.t. the weak-* approximation. (Note that the left-hand side above has this robustness property.) Let us simplify the arguments, and begin, as before, by assuming that both involved measures, that is, $\hat{\mu}$ and $\delta\mu$, are absolutely continuous. Then, the integration by parts is classical. We have (the dependence of function on t is omitted for convenience)

$$\int_{t_0}^{t_1} d\Big(\psi\xi''(\delta y, \delta w)\Big) = \int_{t_0}^{t_1}\Big((-\psi f'_x - \psi G'_x\hat{m})\xi''(\delta y, \delta w) + \psi\frac{d}{dt}\xi''(\delta y, \delta w)\Big)dt. \tag{3.67}$$

Let us compute the full derivative of ξ'' w.r.t. time:

$$\tfrac{d}{dt}\xi''(\delta y, \delta w) = \tfrac{d}{dt}\Big(\xi''_{yy}[\delta y]^2 + 2\xi''_{yw}[\delta y, \delta w] + \xi''_{ww}[\delta w]^2\Big) = \xi'''_{yyy}\eta'_x f[\delta y]^2 +$$
$$\xi'''_{yyw}\hat{m}[\delta y]^2 + 2\xi'''_{ywy}\eta'_x f[\delta y, \delta w] + 2\xi'''_{yww}\hat{m}[\delta y, \delta w] + \xi'''_{wwy}\eta'_x f[\delta w]^2 + \xi'''_{www}\hat{m}[\delta w]^2 +$$
$$2\xi''_{yy}[g'_y\delta y + g'_w\delta w + g'_u\delta u, \delta y] + 2\xi''_{yw}[g'_y\delta y + g'_w\delta w + g'_u\delta u, \delta w] + 2\xi''_{yw}[\delta y, \delta m] +$$
$$2\xi''_{ww}[\delta w, \delta m] = -A_3(\delta y, \delta w) + \xi'''_{yyw}\hat{m}[\delta y]^2 + 2\xi'''_{yww}\hat{m}[\delta y, \delta w] + \xi'''_{www}\hat{m}[\delta w]^2 +$$
$$2\xi''_{yy}[\eta'_x f'_x\xi'_y\delta y, \delta y] - 2\xi''_{yy}[\eta'_x\xi''_{yy}\eta'_x f\delta y, \delta y] + 2\xi''_{yy}[\eta'_x f'_x G\delta w, \delta y] - 2\xi''_{yy}[\eta'_x G'_x f\delta w, \delta y] +$$
$$2\xi''_{yy}[\eta'_x f'_u\delta u, \delta_y] + 2\xi''_{yw}[\eta'_x f'_x\xi'_y\delta y, \delta w] - 2\xi''_{yw}[\eta'_x\xi''_{yy}\eta'_x f\delta y, \delta w] + 2\xi''_{yw}[\eta'_x f'_x G\delta w, \delta w] -$$
$$2\xi''_{yw}[\eta'_x G'_x f\delta w, \delta w] + 2\xi''_{yw}[\eta'_x f'_u\delta u, \delta w] + 2\xi''_{yw}[\delta y, \delta m] + 2\xi''_{ww}[\delta w, \delta m] =$$
$$-A_3(\delta y, \delta w) - A_4(\delta y, \delta w) - A_5(\delta y, \delta w, \delta u) + G''_{xx}[\xi'_y\delta y]^2\hat{m} + G'_x\hat{m}\xi''_{yy}[\delta y]^2 +$$
$$2G''_{xx}[\xi'_y\delta y, \delta w]\hat{m} + 2G'_x\hat{m}G'_x\xi'_y[\delta y, \delta w] + G''_{xx}[G\delta w]^2\hat{m} + G'_x\hat{m}G'_x G[\delta w]^2 +$$
$$2G'_x[\xi'_y\delta y, \delta m] + 2G'_x[G\delta w, \delta m] =$$
$$= -A_3 - A_4 - A_5 + G''_{xx}[\delta x]^2\hat{m} + 2G'_x[\delta x, \delta m] + G'_x\hat{m}\xi''(\delta y, \delta w).$$

Above, (3.64) and the fact that $\xi'_w = G(\xi)$ have been used. Now, by taking into account this expression, and, in view of (3.65), (3.66), and (3.67), derive

$$\tilde{\Omega}_\lambda(\xi, \delta u, \delta\mu) = -\int_{t_0}^{t_1}\frac{\partial^2\tilde{H}(t)}{\partial(y, w, u)^2}[(\delta y, \delta w, \delta u)]^2 dt + \frac{\partial^2\tilde{L}}{\partial z^2}(\hat{z}, \lambda)[\delta z]^2 =$$
$$-\int_{t_0}^{t_1}\psi_y\eta'_x\sum_{i=1}^{5}A_i dt + L''_{pp}(\hat{p}, \lambda)[\delta p]^2 - \int_{t_0}^{t_1} d\Big(\psi\xi''[(\delta y, \delta w)]^2\Big) =$$

$$-\int_{t_0}^{t_1} \psi \sum_{i=1}^{5} A_i dt + L''_{pp}(\hat{p}, \lambda)[\delta p]^2 + \int_{t_0}^{t_1} \psi \cdot \Big(A_2 + G'_x \hat{m} \xi''(\delta y, \delta w) \Big) dt +$$

$$\int_{t_0}^{t_1} \psi \cdot \Big(A_3 + A_4 + A_5 - G''_{xx}[\delta x]^2 \hat{m} - 2G'_x[\delta x, \delta m] - G'_x \hat{m} \xi''(\delta y, \delta w) \Big) dt =$$

$$L''_{pp}(\hat{p}, \lambda)[\delta p]^2 - \int_{t_0}^{t_1} \psi \Big(A_1 + G''_{xx}[\delta x]^2 \hat{m} + 2G'_x[\delta x, \delta m] \Big) dt = \Omega_\lambda(\xi, \delta u, \delta \mu).$$

Here, the fact that $A_1 = f''(\delta x, \delta u)$ has been used. Thereby, $\Omega_\lambda = \tilde{\Omega}_\lambda$ when the measures are absolutely continuous. The general case of arbitrary measures can easily be considered in view of Exercises 3.8 and 3.11 and by virtue of the same idea as in the proof of Proposition 3.1. Note that the value $\tilde{\Omega}_\lambda(\xi, \delta u, \delta \mu)$ does not depend on the method of approximation of $\hat{\mu}$, $\delta \mu$ by absolutely continuous measures. Then, by taking into account the fact that functions $x(t)$, $\psi(t)$, $\delta x(t)$ do not depend on the method of approximation by absolutely continuous measures either, and by arranging an appropriate piecewise linear approximation (as in the proof of Proposition 3.1, see function $\bar{w}_i$ therein), the final formula for Ω_λ, that is (3.53), can be obtained. Now, (3.55) follows from (3.63).

The proof is complete. □

Remark 3.1 The second-order form Ω_λ is robust w.r.t. weak-* perturbations of the measures $\hat{\mu}$ and $\delta \mu$, as are the solutions x, ψ, δx. This important fact follows from the proof of Theorem 3.5. (See also Exercises 3.8, 3.11 in this regard.)

Remark 3.2 With help of a more powerful technique, which is developed in the next chapters (based on the Lebesgue discontinuous variable change), the Frobenius condition can be dispensed with, and second-order optimality conditions, similar to those from Theorem 3.5, can be proved for a general impulsive control problem. This, however, will require the notion of attached family of controls, which will now be described in Chap. 4.

3.7 Exercises

Exercise 3.1 Show that the solution concept (3.4) is not well posed.

Exercise 3.2 Prove the invertibility of the Frobenius theorem. That is, the existence of a local solution implies the symmetry of bilinear mapping $\mathscr{F}$.

Exercise 3.3 Ensure that (H3) is equivalent to the Frobenius condition, that is, to the symmetry of bilinear mapping $\mathscr{F}$, when $G = G(x)$.

Exercise 3.4 Prove the extended version of the Gronwall inequality (Exercise 1.2). That is, for a given scalar nonnegative function $x \in C(T; \mathbb{R})$, such that

$$x(t) \le a + \int_{t_0}^{t} m(\varsigma)x(\varsigma)d\varsigma,$$

where $a \ge 0$, $m(t) \ge 0$, and $m(\cdot)$ is integrable, prove that

$$x(t) \le a \cdot e^{\int_{t_0}^{t} m(\varsigma)d\varsigma}.$$

Exercise 3.5* Prove the following assertion. Given $\mu_i \in C^*(T; \mathbb{R})$, $f_i \in BV(T; \mathbb{R})$, such that $\mu_i \overset{w*}{\to} \mu$, $|\mu_i| \overset{w*}{\to} \nu$, and $\operatorname{Var} f_i|_{[a,b]} \le \lambda_i([a, b])$, where $\lambda_i \in C^*(T; \mathbb{R})$, and $\lambda_i \overset{w*}{\to} \lambda$. Assume that

$$f_i(t) \to f(t) \ \ \forall t \in \mathrm{Cs}(\lambda),$$

where $f \in BV(T; \mathbb{R})$, and $\mathrm{Ds}(\nu) \cap \mathrm{Ds}(\lambda) = \varnothing$. Then,

$$\int_T f_i(t)d\mu_i \to \int_T f(t)d\mu.$$

By constructing an appropriate counterexample, ensure that the above requirement that the total variations of f_i must be majorized by a weakly-* convergent sequence of measures λ_i cannot be omitted.

Exercise 3.6 Prove, solely under (H1), that, if solution $x(\cdot)$ to Eq. (3.2) exists then it is unique. For that, prove and use the following even more extended version (compared with Exercise 3.4) of the Gronwall inequality.

That is, for a given scalar nonnegative function $x \in BV(T; \mathbb{R})$, such that

$$x(t) \le a + \int_{[t_0,t]} x(\varsigma)d\mu,$$

where $a \ge 0$, $\mu \in C^*(T; \mathbb{R})$, is such that $\mu \ge 0$, and $\mathrm{Ds}(\mu) \cap \mathrm{Ds}(x) = \varnothing$, prove that

$$x(t) \le a \cdot e^{\|\mu\|}.$$

(Here, $\mathrm{Ds}(x)$ stands for the set of jumps of $x(\cdot)$.)

Use Exercise 3.5 to prove this. Note that, eventually, the whole line of inequalities with increasing complexity constituted by Exercises 1.2, 3.4, and 3.6 has been built up.

Exercise 3.7 Ensure that the Frobenius condition implies the absolute continuity of the impulsive Hamiltonian $Q^j(t)$, $j = 1, \dots, k$.

Exercise 3.8 Verify that the assertion of Proposition 3.1 also holds true for the conjugate equation (3.15). Thus, the jumps of ψ do not depend on the method of approximation of the measure as with the jumps of state variable x. This fact is somewhat surprising bearing in mind that the Frobenius condition for the joint matrix formed by $G(x,t)$ and $Q'_x(x,\psi,t)$ w.r.t. the variables x, ψ, may not be valid. However, the properties of Hamiltonian systems are important here. Use the proof of Theorem 3.3.

Exercise 3.9 Verify that the subspace X_a of absolutely continuous measures is closed in $X = C^*([0,1];\mathbb{R})$. Let $A : X \to \mathbb{R}^k$ be a weak-* continuous linear operator. Prove that $\operatorname{im} A = A(X_a)$.

Exercise 3.10 Show that $\xi''_{yy}\eta'_x = -\xi'_y\eta''_{xx}\xi'_y$.

Exercise 3.11 Verify that the assertion of Proposition 3.1 also holds true for the variational equation in (3.52). Thus, the jumps of δx do not depend on the method of approximation of the measures $\hat{\mu}$ and $\delta\mu$ by absolutely continuous measures. This exercise is similar to Exercise 3.8. Use the proof of Theorem 3.5.

References

1. Agrachev, A., Gamkrelidze, R.: Index of extremality and quasiextremality. Russian Math. Doklady **284**, 11–14 (1985)
2. Agrachev, A., Gamkrelidze, R.: Quasi-extremality for control systems. J. Sov. Math. **55**(4), 1849–1864 (1991)
3. Arutyunov, A.: Perturbations of extremal problems with constraints and necessary optimality conditions. J. Sov. Math. **54**(6), 1342–1400 (1991)
4. Arutyunov, A.: Second-order conditions in extremal problems. The abnormal points. Trans. Am. Math. Soc. **350**(11), 4341–4365 (1998)
5. Arutyunov, A.: Necessary conditions for an extremum and an inverse function theorem without a priori normality assumptions. Tr. Mat. Inst. Steklova **236** (Differ. Uravn. i Din. Sist.), 33–44 (2002)
6. Arutyunov, A.: Smooth abnormal problems in extremum theory and analysis. Russ. Math. Surv. **67**(3), 403–457 (2012)
7. Arutyunov, A., Dykhta, V., Pereira, F.: Necessary conditions for impulsive nonlinear optimal control problems without a priori normality assumptions. J. Optim. Theory Appl. **124**(1), 55–77 (2005)
8. Arutyunov, A., Jacimovic, V., Pereira, F.: Second order necessary conditions for optimal impulsive control problems. J. Dyn. Control Syst. **9**(1), 131–153 (2003)
9. Arutyunov, A., Karamzin, D.: Necessary conditions for a weak minimum in an optimal control problem with mixed constraints. Differ. Equ. **41**(11), 1532–1543 (2005)
10. Arutyunov, A., Vereshchagina, Y.: On necessary second-order conditions in optimal control problems. Differ. Equ. **38**(11), 1531–1540 (2002)
11. Arutyunov, A.V.: Optimality conditions. Abnormal and degenerate problems. In: Mathematics and its Applications. Kluwer Academic Publishers, Dordrecht (2000)

12. Cartan, H.: Differential Forms. Dover Books on Mathematics. Dover Publications (2012)
13. Dykhta, V.: The variational maximum principle and second-order optimality conditions for impulse processes and singular processes. Sib. Math. J. **35**(1), 65–76 (1994)
14. Dykhta, V., Samsonyuk, O.: Optimal Impulse Control with Applications. Fizmatlit, Moscow (2000) (in Russian)
15. Gamkrelidze, R.: Principles of Optimal Control Theory. Plenum Press, New York (1978)
16. Hartman, P.: Ordinary Differential Equations, 2nd edn. Classics in Applied Mathematics. Society for Industrial and Applied Mathematics, Philadelphia, USA (1982)
17. Hille, E., Phillips, R.: Functional Analysis and Semi Groups. Colloquium Publications, American Mathematical Society (1957)
18. Karamzin, D.: Necessary conditions of the minimum in an impulse optimal control problem. J. Math. Sci. **139**(6), 7087–7150 (2006)
19. Krasnosel'skij, M., Pokrovskij, A.: Systems with Hysteresis. Trans. from the Russian by Marek Niezgódka. Springer, Berlin etc. (1989)
20. Milyutin, A., Osmolovskii, N.: Calculus of variations and Optimal Control. American Mathematical Society, Providence, RI (1998)
21. Sarychev, A.: Optimization of generalized (impulse) controls in a nonlinear time-optimal problem. Diff. Equ. **27**(5), 539–550 (1991)

Chapter 4
Impulsive Control Problems Without the Frobenius Condition

Abstract In this chapter, the same dynamical system and problem as in the previous chapter are considered; however, the Frobenius condition may now be violated. It follows that the violation of the Frobenius condition implies that the constructed extension of the problem is not well posed, since the vector-valued Borel measure in this case may generate an entire integral funnel of various trajectories corresponding to the given dynamical control system with measure. Hence, a considerable expansion of the space of impulsive controls is required. Then, the impulsive control is no longer defined simply by the vector-valued measure, but is already a pair, that is: the vector-valued Borel measure plus the so-called attached family of controls of the conventional type associated with this measure. For this extension of a new type, the same strategy is applied as earlier. Namely, the existence theorem is established, Cauchy-like conditions for well-posedness are indicated, and the maximum principle is proved. The chapter ends with ten exercises.

4.1 Introduction

The investigation of unbounded nonlinear control systems encounters certain difficulties, which stem from the fact that the passage to the weak-* limit in the right hand of the dynamic system is not feasible. This gives rise to the following phenomenon: The space of Borel vector-valued measures does not include the set of all feasible, or achievable, impulsive controls (that is, those controls which are implemented in practice when approximating a given Borel control vector-valued measure by a sequence of absolutely continuous vector-valued measures). Thus, in order to obtain a proper or well-posed statement of the problem, we are forced to extend the notion

---

The original version of this chapter was revised: Belated correction has been incorporated. The correction to this chapter is available at https://doi.org/10.1007/978-3-030-02260-0_8

© Springer Nature Switzerland AG 2019

A. Arutyunov et al., *Optimal Impulsive Control*, Lecture Notes in Control and Information Sciences 477, https://doi.org/10.1007/978-3-030-02260-0_4

of a feasible impulsive control as well as the notion of a feasible trajectory. This is the basic idea for the forthcoming considerations.

There is still a requirement for the control system to be affine w.r.t. the unbounded control parameter v in this chapter, despite the use of the term "nonlinear." At the same time, the control system is essentially nonlinear w.r.t. state position x. Thus, the objective is now the extension of System (0.5) presented in the Introduction. In spite of increasing complexity, there exist various applications for this particular case, for example, in economics. The model of optimizing the costs of advertisement for two goods has the following form (Vidale-Wolfe model [10]):

$$\text{Minimize} \int_0^1 [\alpha_1 u_1(t) + \alpha_2 u_2(t)]dt \;\; \text{subject to: } \alpha_1, \alpha_2 \geq 0, \; \alpha_1 + \alpha_2 = 1, \quad (4.1)$$

$$\begin{cases} \dot{x}_1 = a \cdot (1 - x_1)u_1 - b \cdot x_1, & x_1(0) \in (0, 1), \\ \dot{x}_2 = c \cdot \left(\dfrac{1 - x_2}{x_1}\right) u_2 - d \cdot x_2, & x_2(0) \in (0, x_1(0)), \end{cases} \quad u_1 \geq 0, \quad u_2 \geq 0.$$

Here, u_1, u_2 are the current costs of advertisement of the goods (value per unit of time), x_1, x_2 are the volumes of sale, and $\alpha_1, \alpha_2, a, b, c, d$ are some constants.

It follows that for this problem, the extension procedure provided in Chap. 1 is not well posed, since it leads to a whole funnel of arcs corresponding to a given vector-valued measure. In general, this funnel contains more than a single feasible trajectory. In this chapter, a particular method is proposed to investigate the above phenomenon and class of problems.

The technique of proofs applied throughout the chapter is well known. It is the so-called *method of discontinuous time variable change*. This variable change appears, in multiple forms, in most proofs of this chapter. Historically, the discontinuous time variable change was proposed by H. Lebesgue who found how to reduce the Stieltjes integral to the Riemann integral by using this variable change (see [4], p. 98). In the context of optimal impulsive control, the discontinuous time variable change was first applied in [9, 11], where the authors demonstrated how the impulsive control problem can, in fact, be reduced to a conventional optimal control problem by using this time variable change. The same method sheds light on the nonlinear systems with above-mentioned ambiguity w.r.t. selection of a feasible arc, as in Example (4.1), and uncovers the foundations of the solution concept addressed in the next section.

There is extensive literature on the topic of nonlinear impulsive control systems. Some references are given in the Introduction. In what follows, the approach proposed in [1, 2, 6] is applied. According to this approach, the impulsive control is the pair, that is, the control vector-valued measure supplemented with the so-called attached family of control functions. Let us proceed with the exact formulation.

4.2 Problem Statement and Solution Concept

Consider the following impulsive control problem:

$$
\begin{aligned}
&\text{Minimize} && \varphi(p)\\
&\text{subject to:} && dx = f(x,u,t)dt + G(x,t)d\vartheta,\\
& && x(t_0) = x_0,\;\; x(t_1) = x_1,\;\; t \in T = [t_0, t_1],\\
& && p = (x_0, x_1, t_0, t_1) \in S,\\
& && u(t) \in U \text{ a.a. } t,\\
& && \vartheta = (\mu; \{v_\tau\}),\;\; \text{range}(\mu) \subseteq K.
\end{aligned}
\tag{4.2}
$$

Here, f is a vector-valued function which takes values in the n-dimensional Euclidean space $\mathbb{R}^n$; G is a matrix-valued function with k columns and n rows; $t \in \mathbb{R}^1$ is time, and $T = [t_0, t_1]$, $t_0 < t_1$, is the time interval, which is not fixed; μ is a k-dimensional Borel vector-valued measure defined on T with values in the convex and closed cone K; $\{v_\tau\}$ is a family of measurable vector functions with values in K defined on the interval $[0, 1]$; x is the state variable with values in $\mathbb{R}^n$; the notation "a.a. t" means "for almost all t in the sense of the Lebesgue measure on the real line." The vector u is from $\mathbb{R}^m$. It has the meaning of the usual bounded control parameter. The set U is closed. A feasible control is an essentially bounded measurable function $u(\cdot)$ such that $u(t) \in U$ a.a. t. The pair $(\mu; \{v_\tau\})$ is the impulsive control. The class of impulsive controls is defined below. The vector $p \in \mathbb{R}^{2n+2}$ is called the endpoint vector. The set S is closed. The constraint $p \in S$ is called endpoint. The functions φ, f, and G are continuous.

The meaning of the relation

$$dx = f(x,u,t)dt + G(x,t)d\vartheta \tag{4.3}$$

has not yet been given. Now, let us clarify the definitions of impulsive control and solution to the differential equation (4.3). Let us fix any time interval $T = [t_0, t_1]$.

Consider a Borel vector-valued measure μ such that range$(\mu) \subseteq K$. Let $V(\mu)$ be the set of scalar-valued nonnegative Borel measures ν such that $\exists\, \mu_i$: range$(\mu_i) \subset K$ and $(\mu_i, |\mu_i|) \overset{w*}{\to} (\mu, \nu)$. Here, the symbol $\overset{w*}{\to}$ signifies convergence in the weak-* topology; i.e., all the coordinates of μ_i converge weakly-* in $C^*(T; \mathbb{R})$, and also the variation measure $|\mu_i|$ converges weakly* to ν, where $C^*(T; \mathbb{R})$ is the dual space to $C(T; \mathbb{R})$. For instance, if K is embedded in one of the orthants, then $V(\mu) = \{|\mu|\}$ (single-point set). If K is not a pointed cone, then $V(\mu) = \{\nu \in C^*(T; \mathbb{R}) : \;\nu \geq |\mu|\}$. Let us note that always $|\mu| \in V(\mu)$ (hence, $V(\mu) \neq \varnothing$), and also $\nu \geq |\mu|\;\forall\, \nu \in V(\mu)$. Consider an arbitrary scalar-valued measure $\nu \in V(\mu)$ and a real number $\tau \in T$.

Take any $\tau \in T$. Let a measurable vector-valued function $v_\tau : [0, 1] \to K$ be such that

(i) $\displaystyle\sum_{j=1}^{k} |v_\tau^j(s)| = \nu(\tau)$, a.a. $s \in [0, 1]$;

(ii) $\int_0^1 v_\tau(s)ds = \mu(\tau)$.

The following concepts play a key role in the further considerations.

Definition 4.1 A family of vector-valued functions $\{v_\tau\}$, depending on the real parameter $\tau \in T$, is said to be *attached* to the vector-valued measure μ, provided that there exists a scalar-valued measure $\nu \in V(\mu)$ such that, for every τ, conditions (i) and (ii) are satisfied.

Definition 4.2 A pair $\vartheta = (\mu; \{v_\tau\})$ is said to be an impulsive control in Problem (4.2), provided that μ is a vector-valued measure with values in the cone K such that the family of functions $\{v_\tau\}$ is attached to μ.

Let us define the *variation* of the impulsive control ϑ, as the scalar-valued Borel measure ν which exists by means of the definition of attached family $\{v_\tau\}$. Let us use the same notation $|\vartheta| = \nu$, used for the measures, and, also, the notation $\|\vartheta\| = \nu(T)$.

The definition of attached family immediately implies that $\text{Ds}(\mu) \subseteq \text{Ds}(|\vartheta|)$. Note that the inclusion can be strict. Moreover, $v_\tau = 0$ whenever $\tau \notin \text{Ds}(|\vartheta|)$. Therefore, intrinsically, the set $\{v_\tau\}$ depends on the parameter $\tau \in \text{Ds}(|\vartheta|)$ and contains no more than a countable set of functions different from zero. Also, note that in the case in which the cone K is embedded in the first orthant $\mathbb{R}^k_+$, the definition of the attached family becomes simpler. Indeed, in this case, the functions v_τ are nonnegative. This implies $|\vartheta| = |\mu|$, i.e., the variation of the impulsive control coincides with the variation of the vector-valued measure, which is not true for arbitrary cone K. Then, $v_\tau \neq 0 \Leftrightarrow \tau \in \text{Ds}(|\mu|)$ if $K \subseteq \mathbb{R}^k_+$. The latter assertion is also true for arbitrary pointed cone (see Exercise 4.1).

Let $\tau \in T$, $v \in L_\infty([0, 1]; \mathbb{R}^k)$, and $a \in \mathbb{R}^n$. Assume that function $z(\cdot) = z(\cdot; \tau, a, v)$ defined on the time interval $[0, 1]$ is the solution to the following system:

$$\dot{z}(s) = G(z(s), \tau)v(s), \quad s \in [0, 1], \quad z(0) = a.$$

Take a control u and an impulsive control $\vartheta = (\mu; \{v_\tau\})$ on T. A function $x(\cdot)$ defined on the interval T is called solution to Eq. (4.3) corresponding to the triple (x_0, u, ϑ) if $x(t_0) = x_0$, and

$$\begin{aligned} x(t) = x_0 + \int_{t_0}^{t} f(x, u, \varsigma)d\varsigma + \int_{[t_0, t]} G(x, \varsigma)d\mu_c + \\ + \sum_{\tau \in \text{Ds}(|\vartheta|):\, \tau \le t} \Big(x_\tau(1) - x_\tau(0)\Big) \quad \forall t > t_0, \end{aligned} \tag{4.4}$$

where $x_\tau(\cdot) := z(\cdot; \tau, x(\tau^-), v_\tau)$ and μ_c is the continuous component of measure μ. Note, that the sum in (4.4) is well defined as the atomic set is countable.

The triple (x, u, ϑ), $\vartheta = (\mu, \{v_\tau\})$, considered on the time interval $T = [t_0, t_1]$ is called *control process* if x, u, and ϑ satisfy (4.4) on T. A process is said to be *admissible* if all the constraints of Problem (4.2) are satisfied. An admissible process $(\hat{x}, \hat{u}, \hat{\vartheta})$ considered on the time interval $\hat{T} = [\hat{t}_0, \hat{t}_1]$ is said to be *optimal* if, for any admissible process (x, u, ϑ) considered on $T = [t_0, t_1]$, the inequality $\varphi(\hat{p}) \le \varphi(p)$ holds, where $\hat{p} = (\hat{x}_0, \hat{x}_1, \hat{t}_0, \hat{t}_1)$ and $p = (x_0, x_1, t_0, t_1)$.

For Problem (4.2), the basic interest will be in obtaining conditions for existence of an optimal process and also in obtaining necessary optimality conditions in the form of Pontryagin maximum principle [8].

4.3 Well-Posedness

Under some natural hypothesis, the above-introduced concepts are well posed in the following sense of approximation.

(A) Compactness. Let $\{\mu_i\}$ be a sequence of absolutely continuous vector-valued measures weakly-* converging to a vector-valued measure μ. Let x_i be the absolutely continuous trajectory corresponding to μ_i in accordance with the differential equation in (4.3) in which measure μ is replaced by μ_i (thus, the differential equation with measure becomes, in fact, a conventional one). Then, there exists a set of functions $\{v_\tau\}$ attached to the measure μ and a trajectory $x(\cdot)$ which satisfies (4.4) w.r.t. impulsive control $\vartheta = (\mu, \{v_\tau\})$ and which is a limit point of the sequence $\{x_i\}$. More precisely, after passing to a subsequence, we have

$$x_i(t) \to x(t) \;\; \forall t \in \mathrm{Cs}(|\vartheta|). \tag{4.5}$$

(B) Feasibility. Conversely, for any function $x(\cdot)$, solution in the sense of (4.4), there exists a sequence of absolutely continuous vector-valued measures μ_i weakly-* converging to μ, such that the sequence of the corresponding trajectories x_i converges to x in the sense of (4.5).

The strict formulations and proofs of the presented facts are given later in this chapter. By using (A) and (B), let us briefly comment on Definitions 4.1, and 4.2.

Let us fix a vector-valued measure μ and begin to examine each of all possible families of functions $\{v_\tau\}$ attached to μ, each time constructing a new trajectory $x(\cdot)$ by using Eq. (4.4). As a result, a certain set of solutions is obtained, which is denoted by $\mathscr{F} = \mathscr{F}(\mu)$. It follows from the above statements that, firstly, the constructed set $\mathscr{F}$ involves any approximation method w.r.t. absolutely continuous measures. That is, if, for a trajectory $x(\cdot)$, there exists its absolutely continuous approximation by solutions to (4.4) w.r.t. absolutely continuous measures μ_i such that $\mu_i \xrightarrow{w*} \mu$, then $x \in \mathscr{F}$. Secondly, the set $\mathscr{F}$ does not contain any "extra" solutions, i.e., those solutions for which no such approximations exist. Hence, we come to the conclusion that the set $\mathscr{F}$ is exactly the so-called integral funnel of solutions resulting from the approximation of μ by absolutely continuous measures.

Therefore, generally speaking, arbitrary vector-valued measure μ generates a whole "basket" of trajectories, each of them claiming to be a "solution" to the dif-

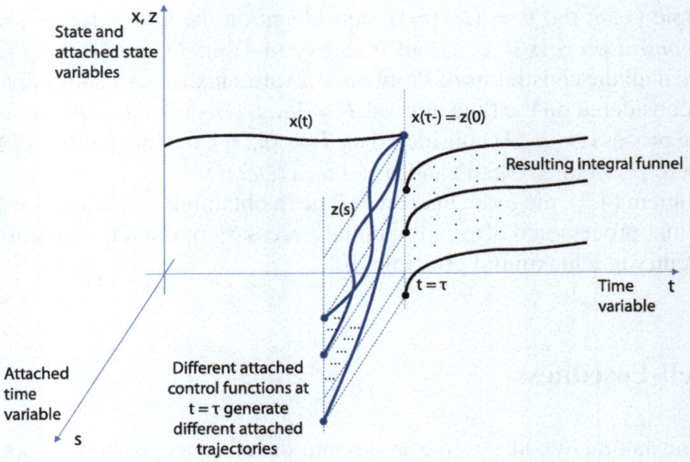

Fig. 4.1 Integral funnel of solutions

ferential equation in (4.3). So, in order to make the concept of solution well posed, it is necessary to extract a selection from this integral funnel. Such a selection is made by introducing the family $\{v_\tau\}$ attached to μ. This situation is shown in Fig. 4.1; see it together with Fig. 3.1.

After fixing a control u, a measure μ, and a family $\{v_\tau\}$ attached to it, according to the triple $(u, \mu, \{v_\tau\})$ and initial value x_0, one can uniquely define the trajectory $x(\cdot)$. As has been illustrated above, the attached family $\{v_\tau\}$ reflects the method of approximation of μ by absolutely continuous measures and, in a certain sense, is nothing but the "interaction scheme of the components of vector measure on the discontinuities of the dynamical system." It becomes clear from what has been said that such an "interaction scheme," together with the vector-valued measure itself, should be included in a joint control parameter. This new control object matches the above-presented concept of impulsive control.

Let us highlight the case of the Frobenius condition considered in Chap. 3. This case is specific but important for applications. Recall that the Frobenius condition is satisfied for System (4.3) if the vector fields G^j are pairwise commutative (here G^j stands for the jth column of matrix G); that is, the symmetry condition (3.5) is satisfied. In much of the work on impulsive control, such an assumption is assumed to be a priori satisfied (see, e.g., [3] and the bibliography cited therein). It follows that, under the Frobenius condition, the above-described integral funnel $\mathscr{F}(\mu)$ of solutions degenerates and becomes just single-valued. That is, the trajectory solution is unique for any vector-valued measure μ.[1] Then, it is clear that the impulsive control could be described using merely the vector-valued measure. Therefore, the introduction of the attached family is redundant (unless state constraints are imposed, see Chap. 5). Thus, under the Frobenius condition, the situation is considerably simpler because the definition of solution becomes the one fully determined solely by vector-valued measure.

[1]See the material of Chap. 3. Consider also Exercise 4.2.

If the Frobenius condition is not satisfied then, in general, the integral funnel $\mathscr{F}(\mu)$ contains "a large number" of trajectories. This is illustrated by the following example.

Example 4.1 Let $n = 1, k = 2, K = \mathbb{R}^2, T = [0, 1]$. Consider the following dynamical system with $\mu = (\mu_1, \mu_2)$:

$$dx = xd\mu_1 + x^2 d\mu_2, \quad x(0) = 1.$$

The Frobenius condition is not satisfied.

Take $\mu = 0$, and show that the integral funnel $\mathscr{F}(0)$ contains at least two different trajectories (in fact, it contains many more trajectories: the entire continuum). One trajectory can be found trivially: It is $x(t) \equiv 1$ corresponding to the zero attached family $v_\tau \equiv 0 \,\forall\, \tau$. Let us construct the second trajectory $\tilde{x}(\cdot)$. We set $v_\tau \equiv 0 \,\forall\, \tau > 0$, $v(0) = 2$, and define the attached controls

$$v_0^1(s) = \begin{cases} 0, & s \in \left[0, \frac{1}{4}\right], \\ -2, & s \in \left(\frac{1}{4}, \frac{1}{2}\right], \\ 1, & s \in \left(\frac{1}{2}, 1\right], \end{cases} \qquad v_0^2(s) = \begin{cases} 2, & s \in \left[0, \frac{1}{4}\right], \\ 0, & s \in \left(\frac{1}{4}, \frac{1}{2}\right], \\ -1, & s \in \left(\frac{1}{2}, 1\right]. \end{cases}$$

Let us find the solution $\alpha(s)$ to the differential system attached to point $\tau = 0$. It is easy to see that

$$\alpha(s) = \frac{1}{1 - 2s}, \quad s \in \left[0, \frac{1}{4}\right] \quad \text{and} \quad \alpha(s) = 2\sqrt{e}e^{-2s}, \quad s \in \left[\frac{1}{4}, \frac{1}{2}\right].$$

Now, note that $\alpha(1/2) = 2/\sqrt{e} \neq 1$, while $\alpha = 1$ is a stationary point of the system on the remaining interval $[1/2, 1]$. This implies $\alpha(1) = c \neq 1$, and hence $\tilde{x}(t) = c$ for $t > 0$; i.e., the trajectory corresponding to the considered attached family is not identically equal to one. This is what was required to show.

Example 4.1 is also interesting as it shows that one can associate a discontinuous trajectory to a nonatomic measure by virtue of (4.4). Surprising as it may seem, this is possible, as cone K is not pointed in Example 4.1. This fact illustrates that true impulsive control in control problems with vector-valued measures represents a broader concept than the one defined solely by the vector-valued measure itself.

Let us introduce some technical constructions which are needed for the proof of the main results of this chapter. Take an arbitrary impulsive control $\vartheta = (\mu, \{v_\tau\})$. Consider the function $\pi : T \to [0, 1]$,

$$\pi(t) = \frac{t - t_0 + |\vartheta|([t_0, t])}{t_1 - t_0 + \|\vartheta\|}, \quad t \in (t_0, t_1], \quad \pi(t_0) = 0, \tag{4.6}$$

which is called discontinuous time variable change (originally proposed by H. Lebesgue). There exists an inverse function $\theta(s)$: $[0, 1] \to T$ such that

(a) $\theta(s)$ is increasing function on [0, 1].
(b) $\theta(s)$ is Lipschitz continuous, that is, $|\theta(s) - \theta(t)| \le \text{const} \cdot |s - t| \,\forall s, t$.
(c) $\theta(s) = \tau, \forall s \in \Gamma_\tau, \forall \tau \in T$, where $\Gamma_\tau = [\pi(\tau^-), \pi(\tau^+)]$.

Here, $c := t_1 - t_0 + |\vartheta|(T)$.

Indeed, when measure $|\vartheta|$ has no atoms, then this fact is obvious from the classical theory. Suppose that $|\vartheta|$ has a finite number of atoms $\tau_1, \ldots, \tau_N \in \text{Ds}(|\vartheta|)$. Then, the required assertion is simple to obtain with the step-by-step application of the continuous time variable change on intervals $[\tau_i, \tau_{i+1}]$. Finally, for arbitrary $|\vartheta|$, the assertion can be proved in view of approximation of $|\vartheta|$ by the sum of its continuous component and a finite linear combination of Dirac's measures.

Let us note that function $\pi(t)$ maps $(\ell + |\vartheta|)$-measurable sets into ℓ-measurable sets. Indeed, it follows directly from the definition of π and from the representation of a measurable set as a union of Borel and zero measure sets. Therefore, if set A is $(\ell + |\vartheta|)$-measurable, then $\theta^{-1}(A)$ is measurable. This implies that $u(\theta(s))$ is measurable, provided that $u(t)$ is $(\ell + |\vartheta|)$-measurable.

However, it is not restrictive to consider in (4.2) that $u(\cdot)$ is $(\ell + |\vartheta|)$-measurable in view of Exercise 4.3. Indeed, the change of values of u on a zero ℓ-measure set does not affect the evolution of the trajectory $x(\cdot)$. Therefore, in what follows, assume that $u(\cdot)$ is $(\ell + |\vartheta|)$-measurable, without making special reference.

Let us prove Property (A) formulated above. We shall provide a more general form of this compactness property.

Lemma 4.1 *Consider a sequence of impulsive controls $\vartheta_i = (\mu_i; \{v_{\tau,i}\})$ such that $\|\vartheta_i\| \le \text{const} \,\forall i$. Let x_i be a solution to the system*

$$dx = f(x, u, t)dt + G(x, t)d\vartheta_i, \quad x(t_0) = x_0, \quad t \in [t_0, t_1],$$

where $x_0, t_0, t_1, u(\cdot)$ are fixed. Suppose that there exists an integrable function $m(t) : T \to \mathbb{R}_+$ such that:

$$|f(x, u, t)| + |G(x, t)| \le m(t)(1 + |x|) \;\; \forall (x, u, t) \in \mathbb{R}^n \times U \times T. \tag{4.7}$$

Then, there exist an impulsive control $\vartheta = (\mu, \{v_\tau\})$ and a trajectory $x(\cdot)$ corresponding to this impulsive control in view of (4.4) such that (4.5) is true after passing to a subsequence.

Proof The proof is given in two stages. Hereinafter, the modulus of a vector signifies the sum of absolute values of its components, that is, $|v| = \sum_i |v^i|$.

Stage 1. Assume that $K \subseteq \mathbb{R}^k_+$ (that is, cone K is embedded in the nonnegative orthant). Consider the following ordinary control differential system:

$$\begin{cases} \dot{x} = \alpha f(x, u, \chi) + G(x, \chi)v, \\ \dot{\chi} = \alpha, \;\; s \in [0, 1], \\ x(0) = x_0, \;\; \chi(0) = t_0, \;\; \chi(1) = t_1, \\ \alpha \ge 0, \;\; v \in K, \; \alpha + |v| = c. \end{cases} \tag{4.8}$$

where control functions $u(s)$, $v(s)$ are ordinary; that is, they are measurable and essentially bounded, while scalar function $\alpha(s)$ and number c are auxiliary control parameters. Let us establish a one-to-one correspondence between the control processes (x, u, ϑ) of System (4.3) defined on time interval T and the control processes $(\tilde{x}, \chi, \tilde{u}, v, \alpha, c)$ of System (4.8) defined on time interval $[0, 1]$.

Firstly, consider a control process (x, u, ϑ) of System (4.3) and the discontinuous time variable change $\pi(t)$ defined in (4.6). Note that function $\pi(t)$ maps time interval T into $[0, 1]$. The corresponding inverse function is denoted by $\theta(s) : [0, 1] \to T$. Function $\theta(s)$ satisfies the properties (a)–(c) mentioned earlier.

Define $c := t_1 - t_0 + |\vartheta|(T)$ and consider $\tilde{u}(s) = u(\theta(s))$,

$$\alpha(s) = \begin{cases} m_1(\theta(s)), & \text{if } s \notin \bigcup_{\tau \in \mathrm{Ds}(|\vartheta|)} \Gamma_\tau, \\ 0, & \text{otherwise}, \end{cases}$$

$$v(s) = \begin{cases} m_2(\theta(s)), & \text{if } s \notin \bigcup_{\tau \in \mathrm{Ds}(|\vartheta|)} \Gamma_\tau, \\ c \cdot \dfrac{v_\tau(\xi_\tau(s))}{\ell(\Gamma_\tau)}, & \text{if } s \in \Gamma_\tau, \end{cases}$$

where function $\xi_\tau(s) = (s - \pi(\tau^-))/\ell(\Gamma_\tau)$ maps $\Gamma_\tau \to [0, 1]$, m_1 and m_2 are the Radon–Nykodim derivatives of measures ℓ and μ_c w.r.t. $\ell + |\vartheta|$ multiplied by the number c, sets Γ_τ are introduced above. Note that $v(s) \in K$ and $\alpha(s) + |v(s)| = c$ for a.a. s since $m_1 + |m_2| = c$.

In view of the definitions, $\int_0^s \alpha(\sigma)d\sigma = \int_{t_0}^{\theta(s)} m_1(\theta(\pi(l)))d\pi(l) = \int_{t_0}^{\theta(s)} d\ell = \theta(s) - t_0$. Therefore, $\chi(\cdot) = \theta(\cdot)$. Now, by applying the discontinuous time variable change we obtain that trajectory $\tilde{x}(\cdot)$ corresponding to $(x_0, \tilde{u}, v, \alpha, c)$ in view of the dynamics in (4.8) equals $x^{\mathrm{ext}}(\cdot)$, where

$$x^{\mathrm{ext}}(s) = \begin{cases} x(\theta(s)), & \text{if } s \notin \bigcup_{\tau \in \mathrm{Ds}(|\vartheta|)} \Gamma_\tau, \\ x_\tau(\xi_\tau(s)), & \text{if } s \in \Gamma_\tau \text{ for some } \tau \in \mathrm{Ds}(|\vartheta|). \end{cases}$$

Indeed, this is due to the chain of equalities for $\sigma = \pi(l)$, and $s \notin \bigcup_{\tau \in \mathrm{Ds}(|\vartheta|)} \Gamma_\tau$:

$$\tilde{x}(s) - x_0 = \int_0^s \alpha(\sigma) f(\tilde{x}(\sigma), \tilde{u}(\sigma), \chi(\sigma))d\sigma + \int_0^s G(\tilde{x}(\sigma), \chi(\sigma))v(\sigma)d\sigma$$

$$= \int_{t_0}^{\theta(s)} m_1(\theta(\pi(l))) f(\tilde{x}(\pi(l)), \tilde{u}(\pi(l)), \theta(\pi(l)))d\pi(l)$$

$$+ \int_{t_0}^{\theta(s)} G(\tilde{x}(\pi(l)), \theta(\pi(l)))m_2(\theta(\pi(l)))d\pi(l)$$

$$+ \sum_{\tau \in \mathrm{Ds}(|\vartheta|),\ \tau < \theta(s)} \int_{\Gamma_\tau} c \cdot G(\tilde{x}(\sigma), \tau) v(\sigma) d\sigma$$

$$= \int_{t_0}^{\theta(s)} m_1(l) f(x(l), u(l), l) d\pi(l)$$

$$+ \int_{t_0}^{\theta(s)} G(x(l), l) m_2(l) d\pi(l)$$

$$+ \sum_{\tau \in \mathrm{Ds}(|\vartheta|),\ \tau < \theta(s)} \int_{\Gamma_\tau} c \cdot G(x_\tau(\xi_\tau(\sigma)), \tau) \frac{v_\tau(\xi_\tau(\sigma))}{\ell(\Gamma_\tau)} d\sigma$$

$$= \int_{t_0}^{\theta(s)} f(x(l), u(l), l) d\ell$$

$$+ \int_{t_0}^{\theta(s)} G(x(l), l) d\mu_c + \sum_{\tau \in \mathrm{Ds}(|\vartheta|),\ \tau < \theta(s)} \Big[x_\tau(1) - x_\tau(0) \Big] = x(\theta(s)) - x_0.$$

Here, we used Property (c) and the definitions of α, v. When $s \in \Gamma_\beta$ for some $\beta \in \mathrm{Ds}(|\vartheta|)$, and thereby $\theta(s) = \beta$, then the extra term $x_\beta(\xi_\beta(s)) - x_\beta(0)$ should be added above. This relation is proved in the same manner as Properties (a)–(c).

Conversely, consider an admissible process $(\tilde{x}, \chi, \tilde{u}, v, \alpha, c)$ in System (4.8). Consider function $\chi(s)$ as the inverse function $\theta(s)$ to some discontinuous time change $\pi : T \to [0, 1]$. The function π is defined uniquely as such a function that $\pi(\chi(s)) = s$, a.a. s: $\alpha(s) > 0$, $\pi(\chi(0)) = 0$, $\pi(\chi(1)) = 1$, and $\pi(t)$ is right-continuous on (t_0, t_1).

Once π is determined, we define the measures μ and $\nu = |\mu|$:

$$F(t, \mu) = \int_0^{\pi(t)} v(s) ds,$$

$$F(t, \nu) = \int_0^{\pi(t)} |v(s)| ds = \int_0^{\pi(t)} (c - \alpha(s)) ds = c\pi(t) - (t - t_0).$$

Consider $u(t) = \tilde{u}(\pi(t))$, $v_\tau(s) = c^{-1}\ell(\Gamma_\tau)v(\gamma_\tau(s))$, where

$$\Gamma_\tau = [\pi(\tau^-), \pi(\tau^+)], \quad \gamma_\tau(s) = \ell(\Gamma_\tau)s + \pi(\tau^-) : [0, 1] \to \Gamma_\tau,$$

and $\vartheta = (\mu; \{v_\tau\})$.

Let $x(\cdot)$ be the solution defined by Formula (4.4) corresponding to x_0, u, ϑ. It follows directly from the change of variable that $x(t) = \tilde{x}(\pi(t))$ $\forall t$. It has been shown that, the two Systems, (4.3) and (4.8), are equivalent while their trajectories are related via the discontinuous time change (4.6).

Now, consider Lebesgue time variable change $\pi_i(t)$ for every impulsive control $\vartheta_i - \pi_i : T \to [0, 1]$, $\pi_i(t) = (t - t_0 + \nu_i([t_0, t])) \cdot c_i^{-1}$, where $c_i = t_1 - t_0 + \nu_i(T)$, $\pi_i(t_0) = 0$, $\nu_i = |\vartheta_i|$. Let, also, $\theta_i(s)$: $[0, 1] \to T$ be the inverse to π_i function satisfying (a)–(c). Denote by $(\tilde{x}_i, \chi_i, \tilde{u}_i, v_i, \alpha_i, c_i)$ the process of System (4.8) corresponding to (x_i, u_i, ϑ_i) in view of variable change π_i and by means of the same formulas as above. (Note that in view of Exercise 4.3, it is not restrictive to consider that $u(\cdot)$ is $(\ell + \nu_i)$-measurable for all i. So, $\tilde{u}_i := u \circ \theta_i$ is measurable.)

By the assumption of the lemma, we have the uniform estimate $\|v_i\| \leq \text{const}$. Then, $c_i \leq \text{const} < +\infty$ $\forall i$. Therefore, in view of (4.7), and the Gronwall inequality, the trajectories $\tilde{x}_i$ and $\chi_i(s) = \theta_i(s)$ of System (4.8) are uniformly bounded. Then, by Arzela–Ascoli theorem, after extracting a subsequence, we have $\tilde{x}_i(s) \rightrightarrows \tilde{x}(s)$, $\chi_i(s) \rightrightarrows \chi(s) = \theta(s)$. By using the weak sequential compactness of the unit ball in Hilbert space, and by passing to a subsequence, we have $\alpha_i \xrightarrow{w} \alpha$, $v_i \xrightarrow{w} v$. It is also not restrictive to consider that $c_i \to c$, where $c > 0$. Since K is convex and embedded in the first orthant by assumption, then, in view of weak convergence, $\alpha + |v| = c$.

Note that function $\theta = \chi$ is the inverse function to some discontinuous time variable change π. Since $\tilde{u}_i(s) = u(\theta_i(s))$, then, in view of Exercise 4.4,

$$f(\tilde{x}_i, \tilde{u}_i, \chi_i) \to f(\tilde{x}, \tilde{u}, \theta)$$

in the L_1-norm, where the function $\tilde{u}(s) = u(\theta(s))$ is measurable by the convention adopted above (see Exercise 4.3). By using this fact, passing to the limit in (4.8) we arrive at the conclusion that the obtained process $(\tilde{x}, \theta, \tilde{u}, v, \alpha, c)$ satisfies System (4.8).

Therefore, due to the relations established above, there exist an impulsive control $\vartheta = (\mu; \{v_\tau\})$, and a trajectory $x(t)$ such that $x(t) = \tilde{x}(\pi(t))$ $\forall t \in T$, which satisfy (4.4). Let us confirm that, for this trajectory, Condition (4.5) is valid. By construction, $x_i(t) = \tilde{x}_i(\pi_i(t))$ $\forall t \in T$, and $\tilde{x}_i \rightrightarrows \tilde{x}$. Therefore, in order to prove (4.5), it is obviously sufficient to demonstrate that $\pi_i(t) \to \pi(t)$ $\forall t \in \text{Cs}(|\vartheta|)$.

Let set C be the intersection of $\text{Cs}(|\vartheta|)$ and $\bigcap_i \text{Cs}(|\vartheta_i|)$. It is clear that set C has full measure. Let $t \in C \cap (t_0, t_1)$. Then, $\pi(\theta(s)) = s$, where $s : t = \theta(s)$, and $\pi_i(\theta_i(s_i)) = s_i$, where $s_i : t = \theta_i(s_i)$, for all i. For any limit point $\bar{s}$ of $\{s_i\}$, in view of $\theta_i \rightrightarrows \theta$, one has $\theta(\bar{s}) = \theta(s)$. However, $\theta(s) \in \text{Cs}(|\vartheta|)$, whence $\bar{s} = s$. Therefore, $s_i \to s$, and

$$\pi_i(t) = \pi_i(\theta_i(s_i)) = s_i \to s = \pi(\theta(s)) = \pi(t).$$

We have proved that $\pi_i \to \pi$ on a full measure set. Then, $\pi_i(t) \to \pi(t)\ \forall \mathrm{Cs}(|\vartheta|) \cap (t_0, t_1)$ in view of monotonicity. Convergence at the endpoints t_0, t_1 is valid in view of the definitions of π, π_i: $\pi(t_0) = \pi_i(t_0) = 0$ while $\pi(t_1) = \pi_i(t_1) = 1$.

We conclude that (4.5) is valid.

Stage 2. There exists a standard procedure in control theory that allows a reduction of feasible cone K to a pointed cone embedded in $\mathbb{R}^k_+$ by doubling the dimension of control. Let us describe it.

Along with the control system (4.3), consider the following system

$$\begin{aligned} dx &= f(x,u,t)dt + G(x,t)Pd\bar{\vartheta}, \\ \bar{\vartheta} &= (\bar{\mu}; \{\bar{v}_\tau\}),\ \mathrm{range}(\bar{\mu}) \subseteq \bar{K}. \end{aligned} \tag{4.9}$$

Here, $P : \mathbb{R}^{2k} \to \mathbb{R}^k$ is a linear operator defined by $(P\xi)^j = \xi^{2j} - \xi^{2j-1}$, $j = 1, ..., k$, vectors $\bar{\mu}$ and $\bar{v}_\tau$ have $2k$ components, and

$$\bar{K} = \{\bar{v} \in \mathbb{R}^{2k}_+ : P\bar{v} \in K\}$$

is a closed and convex cone embedded in first orthant.

Note that any Borel measure μ can be represented as $\mu = \mu^+ - \mu^-$, where μ^+, μ^- are nonnegative Borel measures, and $|\mu| = |\mu^+| + |\mu^-|$ (the so-called Jordan decomposition). Any scalar function $v(t)$ can be represented as the difference of two nonnegative functions, i.e., $v = v^+ - v^-$, where $v^+(t) = \max\{v(t), 0\}$, and $v^-(t) = -\min\{v(t), 0\}$. Moreover, $|v(t)| \equiv v^+(t) + v^-(t)$.

By using these relations, for a given impulsive control $\vartheta = (\mu; \{v_\tau\})$ in (4.3), we construct impulsive control $\bar{\vartheta}$ for (4.9) by the formulas

$$\bar{\mu}^{2j-1} := \mu_c^{j-} + \sum_{\tau \in \mathrm{Ds}(|\vartheta|)} \|v_\tau^{j-}\|_{\mathrm{L}_1} \cdot \delta_\tau, \quad \bar{\mu}^{2j} := \mu_c^{j+} + \sum_{\tau \in \mathrm{Ds}(|\vartheta|)} \|v_\tau^{j+}\|_{\mathrm{L}_1} \cdot \delta_\tau,$$

and $\bar{v}_\tau^{2j-1} := v_\tau^{j-}$, $\bar{v}_\tau^{2j} := v_\tau^{j+}$, $j = 1, ..., k$. It is clear that the pair $\bar{\vartheta} = (\bar{\mu}; \{\bar{v}_\tau\})$ satisfies Definition 4.2, while the trajectory generated by this control w.r.t. (4.9) is the same as the one generated by ϑ w.r.t. (4.3). Moreover, $\|\bar{\vartheta}\| = \|\vartheta\|$.

Conversely for a given impulsive control $\bar{\vartheta} = (\bar{\mu}; \{\bar{v}_\tau\})$ in (4.9), we define $\mu = P\bar{\mu}$, and $v_\tau = P\bar{v}_\tau$. Note that $(\mu; \{v_\tau\})$ may not satisfy Definition 4.2 because of Property (i) violated therein. Indeed, the difference of two nonnegative functions may vanish, for example. However, thanks to Exercise 4.5, there exists an attached family $\{\tilde{v}_\tau\}$ such that the pair $\vartheta = (\mu; \{\tilde{v}_\tau\})$ is impulsive control for (4.3), with $\|\vartheta\| \leq \|\bar{\vartheta}\|$, generating just the same trajectory as $\bar{\vartheta}$.

We have established the equivalence of control systems (4.3) and (4.9). The map of correspondence for this equivalence is ambiguous as a single impulsive control of (4.3) gives rise to a whole family of impulsive controls in (4.9), though generating the same trajectory. However, there is unique impulsive control in this family with the same norm of preimage (norms of others are greater than the norm of preimage). In this sense, the equivalence is one-to-one and isometric.

Now, the application of Stage 1 completes the proof. □

Remark 4.1 The assertion of Lemma 4.1 is somewhat simple to strengthen. Namely, one can state that, after passing to a subsequence, (4.5) is valid, and moreover, $\mu_i \overset{w*}{\to} \mu$. Indeed, this follows from the application of Lemma 4.1 to the extended differential system (4.3) with the added trivial extra variables $dy = d\vartheta$.

Property (A) follows immediately from Lemma 4.1 and Remark 4.1. Indeed, by using the uniform boundedness principle, due to the weak-* convergence, we have that $\|\mu_i\| \leq \text{const}$. Then, the assumptions of Lemma 4.1 reinforced with Remark 4.1 are satisfied. So, (4.5) is true.

Property (B) is outlined in the following assertion.

Lemma 4.2 *Under (4.7), for any trajectory $x(t)$ which is a solution to (4.4), corresponding to the impulsive control $(\mu; \{v_\tau\})$ there exists a sequence of absolutely continuous vector-valued measures μ_i weakly-* converging to μ in $C^*(T; \mathbb{R}^k)$ such that the sequence of associated with μ_i trajectories $x_i(t)$ converges to $x(t)$ in the sense of (4.5).*

The proof is straightforward and represents a simple exercise as set out in Exercise 4.6.

The following lemma is a simple generalization of the classical results regarding the well-posedness in the Cauchy sense.

Lemma 4.3 *Let $x_0 \in \mathbb{R}^n$, and (u, ϑ) be control in (4.3). Suppose that (4.7) is in force.*

Then, on the interval T, there exists a solution $x(\cdot)$ to (4.3), corresponding to the control (u, ϑ), and to the initial value $x(t_0) = x_0$.

In the case that Condition (4.7) does not hold, but f, g are continuously differentiable w.r.t. x, and $t_0 \notin Ds(|\vartheta|)$, there exists a number $\delta > 0$ such that the solution $x(\cdot)$ exists only locally on the interval $[t_0, t_0 + \delta]$.

In both cases, the solution is unique.

Moreover, small perturbations of the initial value x_0 produce small deviations of the solution

$$|x(t) - \tilde{x}(t)| \leq const|x_0 - \tilde{x}_0| \quad \forall t \in T,$$

where $\tilde{x}(t)$ is the solution to (4.3) corresponding to the starting value $\tilde{x}_0$, and the same control (u, ϑ).

The proof simply repeats the arguments of Lemma 4.1. In view of the established equivalence of the control systems (4.3) and (4.8), it is sufficient to apply the classical well-posedness results to the conventional control system (4.8).

If, additionally, Lipschitz continuity w.r.t. u is assumed, then a similar estimate is valid, ensuring sensitivity to small perturbations in L_1-norm of the ordinary control

$$|x(t) - \tilde{x}(t)| \leq \text{const}\Big(|x_0 - \tilde{x}_0| + \|u - \tilde{u}\|_{L_1}\Big) \quad \forall t \in T.$$

However, this type of estimate fails as soon as the impulsive control ϑ is perturbed in the weak-* topology.

4.4 Existence of Solution

Once the extension is constructed, one of the key points is the question of existence of solution to the extended problem. Indeed, the extension procedure must be proper or well posed in the sense that the relaxed problem possesses a solution, the so-called generalized solution. The following theorem is a version of the A. F. Filippov existence theorem, [5], though adjusted for discontinuous trajectories.

Theorem 4.1 *Assume that:*

(a) *The sets U and S are compact.*
(b) *The set $f(x, U, t)$ and cone K are convex.*
(c) *Estimate (4.7) is satisfied.*
(d) *There exists an admissible process $(\bar{x}, \bar{u}, \bar{\vartheta})$ given on $\bar{T} = [\bar{t}_0, \bar{t}_1]$ in Problem (4.2).*
(e) *There exists a constant $\kappa > 0$ such that, for any admissible process (x, u, ϑ) on $T = [t_0, t_1]$ such that $\varphi(p) \leq \varphi(\bar{p})$, where $\bar{p} = (\bar{x}(\bar{t}_0), \bar{x}(\bar{t}_1), \bar{t}_0, \bar{t}_1)$, it follows that $\|\vartheta\| \leq \kappa$.*

Then, there exists a solution to Problem (4.2).

Proof It is not restrictive to consider that the estimate $\|\vartheta\| \leq \kappa$ holds over all admissible processes (x, u, ϑ). Indeed, this conclusion is simple to reach by imposing the additional endpoint constraint $\varphi(p) \leq \varphi(\bar{p})$. It is also clear that in view of Reduction (4.9), it is enough to give the proof for the case when $K \subseteq \mathbb{R}^k_+$. The next arguments use the idea underlying the proof of Lemma 4.1.

Consider an auxiliary control problem

$$\begin{aligned} &\text{Minimize} \quad \varphi(p) \\ &\text{subject to:} \quad \dot{x} = \alpha f(x, u, \chi) + (1 - \alpha) G(x, \chi) v, \\ &\qquad \dot{y} = 1 - \alpha, \\ &\qquad \dot{\chi} = \alpha, \;\; s \in [0, s_1], \\ &\qquad p = (x_0, x_1, \chi_0, \chi_1) \in S, \\ &\qquad y_0 = 0, \;\; y_1 \leq \kappa, \\ &\qquad u \in U, \;\; \alpha \in [0, 1], \;\; v \in K, \;\; |v| = 1. \end{aligned} \tag{4.10}$$

Problem (4.10) is considered on the free time interval $[0, s_1]$. Problem (4.10) possesses three control functions $u(s)$, $v(s)$, $\alpha(s)$, which are of the conventional type; i.e., they are measurable and essentially bounded. Thereby, Problem (4.10) is a conventional autonomous control problem with free time.

Let us show that the two Problems, (4.2) and (4.10), are equivalent in the sense that for each admissible process $(x, u, \vartheta, t_0, t_1)$ of problem (4.2), there exists an admissible process $(\tilde{x}, y, \chi, \tilde{u}, v, \alpha, s_1)$ of problem (4.10), such that $\varphi(p) = \varphi(\tilde{p})$, where $p = (x_0, x_1, t_0, t_1)$, and $\tilde{p} = (\tilde{x}_0, \tilde{x}_1, \chi_0, \chi_1)$, and vice versa.

First, take an admissible process $(x, u, \vartheta, t_0, t_1)$ of Problem (4.2). Consider the discontinuous time variable change

$$\pi(t) = t - t_0 + |\vartheta|([t_0, t]) \text{ as } t > t_0, \quad \pi(t_0) = 0.$$

Note that the function $\pi(t)$ maps segment T into $[0, s_1]$, where

$$s_1 = t_1 - t_0 + |\vartheta|(T).$$

The inverse function is denoted by $\theta(s) : [0, s_1] \to T$.

By changing values of $u(t)$ on a zero ℓ-measure set, we ensure that $u(t)$ is measurable w.r.t. the measure $\ell + |\vartheta|$. Such a change of values of function $u(\cdot)$ is not restrictive.

Let us take $\tilde{u}(s) = u(\theta(s))$,

$$\alpha(s) = \begin{cases} m_1(\theta(s)), & \text{if } s \notin \bigcup_{\tau \in \mathrm{Ds}(\vartheta)} \Gamma_\tau, \\ 0, & \text{otherwise}, \end{cases}$$

$$v(s) = \begin{cases} m_2(\theta(s))/(1 - \alpha(s)), & \text{if } s \notin \bigcup_{\tau \in \mathrm{Ds}(\vartheta)} \Gamma_\tau \text{ and } \alpha(s) < 1, \\ (\ell(\Gamma_\tau))^{-1} v_\tau(\xi_\tau(s)), & \text{if } s \in \Gamma_\tau, \end{cases}$$

where m_1 and m_2 are the Radon-Nikodym derivatives of the measures ℓ and μ_c with respect to $\ell + |\vartheta|$, $\Gamma_\tau = [\pi(\tau^-), \pi(\tau^+)]$, $\xi_\tau(s) := \frac{s - \pi(\tau^-)}{\ell(\Gamma_\tau)} : \Gamma_\tau \to [0, 1]$. When $\alpha(s) = 1$, the values of $v(s)$ may be taken as arbitrary unit vectors from K, since the trajectory $x(\cdot)$ does not depend on those values. Note that $v(s) \in K$, $|v(s)| = 1$ due to $m_1(t) + |m_2(t)| = 1$ a.a. $t \notin \mathrm{Ds}(|\vartheta|)$ w.r.t. the measure $\ell + |\vartheta|$.

In view of the definition, $\chi(s) = \theta(s)\ \forall s \in [0, 1]$. By performing the variable change in (4.4), it is simple to see that the trajectory $\tilde{x}(\cdot)$, which is the solution to the dynamical system (4.10) corresponding to the above-constructed quintuple $(x_0, \tilde{u}, v, \alpha, s_1)$, exists and equals the function (see the proof of Lemma 4.1)

$$x^{\mathrm{ext}}(s) = \begin{cases} x(\theta(s)), & \text{if } s \notin \bigcup_{\tau \in \mathrm{Ds}(|\vartheta|)} \Gamma_\tau, \\ x_\tau(\xi_\tau(s)), & \text{if } s \in \Gamma_\tau \text{ for some } \tau \in \mathrm{Ds}(|\vartheta|). \end{cases}$$

Then, $\tilde{x}(s_1) = x(t_1)$. Therefore, $\varphi(p) = \varphi(\tilde{p})$ and $\tilde{p} \in S$. By construction, we also obtain that $y(s_1) = |\vartheta|(T)$. Then, due to (e), the above-constructed process is admissible in (4.10) while the minimizing functional preserves the same value $\varphi(p)$.

Conversely, consider an admissible process $(\tilde{x}, y, \chi, \tilde{u}, v, \alpha, s_1)$ of problem (4.10). The function $\chi(\cdot)$ is the inverse to some discontinuous time variable change $\pi : T \to [0, s_1]$. The function π is uniquely defined as the one which satisfies $\pi(\chi(s)) = s$, a.a. s such that $\alpha(s) > 0$, $\pi(t_0) = 0$, $\pi(t_1) = s_1$, and $\pi(t)$ is right-continuous in (t_0, t_1). Once π is determined, we find the measure μ via its distribution function:

$$F(t,\mu)=\int_0^{\pi(t)}(1-\alpha(s))v(s)ds.$$

Take $u(t)=\tilde{u}(\pi(t))$, $v_\tau(s)=\ell(\Gamma_\tau)v(\gamma_\tau(s))$, where

$$\gamma_\tau(s)=\ell(\Gamma_\tau)s+\pi(\tau^-):[0,1]\to\Gamma_\tau.$$

Let $x(\cdot)$ be the solution defined by (4.4). It follows directly from the change of variable that $x(t_1)=\tilde{x}(s_1)$, $\|\vartheta\|=\|\mu\|=y(s_1)$. It is clear that the endpoint constraints are satisfied. So, the constructed process is admissible in Problem (4.2), while the minimizing functional preserves the same value $\varphi(\tilde{p})$.

Thus, we have demonstrated that the two Problems, (4.2) and (4.10), are equivalent. Therefore, if a solution exists to one of these problems, then it also exists to the other.

Let us ensure the existence of solution to the auxiliary problem (4.10). Indeed, the velocity set in (4.10) is convex due to Condition (b). This, together with other Conditions (a), (c), (d), allows us to apply the classic Filippov existence theorem to Problem (4.10). Therefore, the solution to the auxiliary problem exists. So, it also exists to original Problem (4.2).

Above, implicit use of the fact that the endpoint constraints in (4.10) and Condition (a) imply $s_1\le$ const, is made. Indeed, for a trajectory y of (4.10), we have:

$$\kappa\ge y(s_1)=\int_0^{s_1}(1-\alpha)ds=s_1-(\chi_1-\chi_0)\overset{\text{a)}}{\Longrightarrow}s_1\le\text{const}.$$

Assumption (e) is already implemented in (4.10) with above constant κ.

The proof is complete. □

4.5 Maximum Principle

In this section, assume that all functions φ, f, G are continuously differentiable. Denote by H the Hamilton–Pontryagin function

$$H(x,u,\psi,t):=\langle f(x,u,t),\psi\rangle,$$

and by Q the following vector-valued function

$$Q(x,\psi,t):=G^*(x,t)\psi.$$

Theorem 4.2 *Let the triple* $(\hat{x},\hat{u},\hat{\vartheta})$, $\hat{\vartheta}=(\hat{\mu},\{\hat{v}_\tau\})$, *considered on the time interval* $\hat{T}=[\hat{t}_0,\hat{t}_1]$, *be an optimal process in Problem (4.2). Then, there exist a number* $\lambda\ge 0$

as well as a vector-valued function of bounded variation ψ *defined on* $\hat{T}$*, and for every point* $\tau \in Ds(|\hat{\vartheta}|)$*, there exists an absolutely continuous vector-valued function* ψ_τ *defined on* $[0, 1]$*, such that*

$$\begin{aligned} &\lambda + |\psi(t)| \neq 0 \ \ \forall t \in \hat{T}, \\ &\lambda + |\psi_\tau(s)| \neq 0 \ \ \forall s \in [0, 1] \ \ \forall \tau \in Ds(|\hat{\vartheta}|), \end{aligned} \tag{4.11}$$

$$\psi(t) = \psi(\hat{t}_0) - \int\limits_{\hat{t}_0}^{t} \frac{\partial H}{\partial x}(\varsigma)d\varsigma - \int\limits_{[\hat{t}_0,t]} \frac{\partial}{\partial x}\Big\langle Q(\varsigma), d\hat{\mu}_c\Big\rangle + \sum_{\tau \in Ds(|\hat{\vartheta}|):\, \tau \le t} \Big(\psi_\tau(1) - \psi_\tau(0)\Big) \ \ \forall t \in (\hat{t}_0, \hat{t}_1], \tag{4.12}$$

$$\begin{cases} \dot{\psi}_\tau(s) = -\frac{\partial}{\partial x}\big\langle Q_\tau(s), \hat{v}_\tau(s)\big\rangle, \\ \psi_\tau(0) = \psi(\tau^-), \ \ s \in [0, 1], \end{cases} \tag{4.13}$$

$$(\psi(\hat{t}_0), -\psi(\hat{t}_1), -\phi(\hat{t}_0), \phi(\hat{t}_1)) \in \lambda\varphi'(\hat{p}) + N_S(\hat{p}), \tag{4.14}$$

$$\max_{u \in U} H(u, t) = H(t) = \phi(t), \ \ a.a.\ t \in \hat{T}, \tag{4.15}$$

$$\begin{cases} Q(t) \in K^\circ \ \ \forall t \in \hat{T}, \\ Q_\tau(s) \in K^\circ \ \ \forall s \in [0, 1], \ \forall \tau \in Ds(|\hat{\vartheta}|), \\ \int\limits_{\hat{T}} \Big\langle Q(t), d\hat{\vartheta}\Big\rangle = 0, \end{cases} \tag{4.16}$$

where the function ϕ *is defined* $\forall t \in (\hat{t}_0, \hat{t}_1]$ *as follows*

$$\phi(t) = \phi(\hat{t}_0) + \int\limits_{\hat{t}_0}^{t} \frac{\partial H}{\partial t}(\varsigma)d\varsigma + \int\limits_{[\hat{t}_0,t]} \frac{\partial}{\partial t}\Big\langle Q(\varsigma), d\hat{\mu}_c\Big\rangle + \sum_{\tau \in Ds(|\hat{\vartheta}|):\, \tau \le t} \Big(\phi_\tau(1) - \phi_\tau(0)\Big),$$

$$\begin{cases} \dot{\phi}_\tau(s) = \dfrac{\partial}{\partial t}\big\langle Q_\tau(s), \hat{v}_\tau(s)\big\rangle, \ \ s \in [0, 1], \\ \phi_\tau(0) = \phi(\tau^-). \end{cases}$$

Here, $\hat{p} = (\hat{x}(t_0), \hat{x}(t_1), \hat{t}_0, \hat{t}_1)$, $K^\circ = N_K(0)$ is the polar cone. As usual, if some arguments of H, Q are omitted, then it means that the values $\hat{x}(t)$, $\hat{u}(t)$, or $\psi(t)$ substitute the omitted arguments, e.g., $H(t) = H(\hat{x}(t), \hat{u}(t), \psi(t), t)$. The subindex τ means that the attached values $\hat{x}_\tau(s)$, $\psi_\tau(s)$ at the point of discontinuity τ substitute the omitted arguments: $Q_\tau(s) = Q(\hat{x}_\tau(s), \psi_\tau(s), \tau)$. The same notation is used for the partial derivatives of H, Q. The integral in (4.16) is not a Lebesgue integral, but it is understood in the following sense, analogous to the solution concept (4.4), that is

$$\int_{\hat{T}} \left\langle Q(t), d\hat{\vartheta} \right\rangle = \int_{\hat{T}} \left\langle Q(t), d\hat{\mu}_c \right\rangle + \sum_{\tau \in \mathrm{Ds}(|\hat{\vartheta}|)} \int_{[0,1]} \left\langle Q_\tau(s), \hat{v}_\tau(s) \right\rangle ds.$$

Let us make some brief comments on these conditions. Condition (4.11) is the nontriviality condition which, now, takes this extended form. Condition (4.12) is the equation for the adjoint function ψ which may have jumps. Its jumps are computed with the help of the adjoint conjugate system presented in (4.13), where function ψ_τ plays the role of the adjoint variable along the jump evolution. Conditions (4.14) and (4.15) are conventional. They represent, respectively, the transversality and the maximum condition for the regular component f of the dynamics. Condition (4.16) is an analogue of the maximum condition (4.15), but for the impulsive component G of the dynamics. In particular, note that the impulsive Hamiltionians Q^j can now exhibit jumps, which is not the case in Chap. 3, where Q^j are continuous.

Proof Let us exploit the same idea as in Lemma 4.1 based on the discontinuous change of the time variable (4.6). Let us define $\hat{c} = \hat{t}_1 - \hat{t}_0 + \hat{v}(T)$ and consider the auxiliary problem

$$\begin{aligned} \text{Minimize } & \varphi(p) \\ \text{subject to: } & \dot{x} = \alpha f(x, u, \chi) + (\hat{c} - \alpha) G(x, \chi) v, \\ & \dot{\chi} = \alpha, \; s \in [0, s_1], \\ & p = (x_0, x_1, \chi_0, \chi_1) \in S, \\ & \chi_0 = \chi(0), \; \chi_1 = \chi(s_1), \\ & u \in U, \; \alpha \in [0, \hat{c}], \\ & v \in K, \; |v| = 1, \end{aligned} \tag{4.17}$$

where the controls $u(s)$, $v(s)$ are conventional; i.e., they are measurable and essentially bounded, and scalar function $\alpha(s)$ is an auxiliary control function. The number s_1 is not fixed a priori here, so Problem (4.17) is considered on the nonfixed time interval $[0, s_1]$, and thus, Problem (4.17) is the so-called autonomous problem with free time. Let us show that Problems (4.2) and (4.17) are equivalent in the sense that for every admissible process (x, u, ϑ) of Problem (4.2) considered on $T = [t_0, t_1]$, there exists an admissible process $(\tilde{x}, \chi, \tilde{u}, v, \alpha, s_1)$ of Problem (4.17), such that $\varphi(p) = \varphi(\tilde{p})$, where $\tilde{p} = (\tilde{x}_0, \tilde{x}_1, \chi_0, \chi_1)$, and vice versa.

First, on $T = [t_0, t_1]$, consider an admissible process (x, u, ϑ) of Problem (4.2). Consider the discontinuous time variable change

$$\pi(t) = \frac{t - t_0 + |\vartheta|([t_0, t])}{\hat{c}} \quad \text{as } t > t_0, \quad \pi(t_0) = 0.$$

Note that function $\pi(t)$ maps segment T into $[0, s_1]$, where

$$s_1 = \frac{t_1 - t_0 + |\vartheta|(T)}{\hat{c}}.$$

The corresponding inverse function is denoted by $\theta(s) : [0, s_1] \to T$. Function $\theta(s)$ has the same properties (a), (b), (c) as stated earlier.

Take $\tilde{u}(s) = u(\theta(s))$, $\chi(0) = t_0$,

$$\alpha(s) = \begin{cases} m_1(\theta(s)), & \text{if } s \notin \bigcup_{\tau \in \mathrm{Ds}(|\vartheta|)} \Gamma_\tau, \\ 0, & \text{otherwise,} \end{cases}$$

$$v(s) = \begin{cases} m_2(\theta(s))/(\hat{c} - \alpha(s)), & \text{if } s \notin \bigcup_{\tau \in \mathrm{Ds}(|\vartheta|)} \Gamma_\tau \text{ and } \alpha(s) < \hat{c}, \\ \ell^{-1}(\Gamma_\tau) v_\tau(\xi_\tau(s)), & \text{if } s \in \Gamma_\tau, \end{cases}$$

where m_1, m_2 are the Radon–Nykodim derivatives of measures ℓ and μ_c w.r.t. $\ell + |\vartheta|$ multiplied by the number $\hat{c}$, $\Gamma_\tau = [\pi(\tau^-), \pi(\tau^+)]$, and $\xi_\tau(s) = (s - \pi(\tau^-))/\ell(\Gamma_\tau) : \Gamma_\tau \to [0, 1]$—the scale function. As $\alpha(s) = \hat{c}$, the values of $v(s)$ may be chosen as arbitrary unit vectors as these values do not affect the dynamics in this case.

Note that $v(s) \in K$ and $|v(s)| = 1$ due to $m_1 + |m_2| = \hat{c}$.

By definition, we have $\chi(\cdot) = \theta(\cdot)$. The trajectory $\tilde{x}(\cdot)$ corresponding to $(x_0, \tilde{u}, v, \alpha, s_1)$ in view of the dynamics in (4.17) equals $x^{\text{ext}}(\cdot)$. (See Lemma 4.1). So, $\varphi(p) = \varphi(\tilde{p})$.

Conversely, consider an admissible process $(\tilde{x}, \chi, \tilde{u}, v, \alpha, s_1)$ of Problem (4.17). Take $t_0 = \chi_0$, $t_1 = \chi_1$. We consider function $\chi(s) = \theta(s)$ as the inverse function to some discontinuous time change $\pi : T \to [0, s_1]$, where $T = [t_0, t_1]$. The function π is defined uniquely as such a function that $\pi(\chi(s)) = s$, a.a. s: $\alpha(s) > 0$, $\pi(t_0) = 0$, $\pi(t_1) = s_1$, and $\pi(t)$ is right-continuous in (t_0, t_1).

Once π is determined, we define the measures μ, ν:

$$F(t, \mu) = \int_0^{\pi(t)} (\hat{c} - \alpha(s)) v(s) ds,$$

$$F(t, \nu) = \hat{c}\pi(t) - (t - t_0).$$

Define $u(t) = \tilde{u}(\pi(t))$, $v_\tau(s) = \ell(\Gamma_\tau) v(\gamma_\tau(s))$, where $\gamma_\tau(s) = \ell(\Gamma_\tau)s + \pi(\tau^-) : [0, 1] \to \Gamma_\tau$, $\vartheta = (\mu; \{v_\tau\})$ with $\|\vartheta\| = \nu$. Let $x(\cdot)$ be the solution defined by Formula (4.4). It follows from the variable change that $x(t) = \tilde{x}(\pi(t))$ $\forall t \in T$. Then $p = \tilde{p}$, where $p = (x_0, x_1, t_0, t_1)$, $\tilde{p} = (\tilde{x}_0, \tilde{x}_1, \chi_0, \chi_1)$. Thus, we have constructed the process (x, u, ϑ) satisfying all the constraints of Problem (4.2) over T, and such that $\varphi(p) = \varphi(\tilde{p})$.

We have shown that Problems (4.2) and (4.17) are equivalent. Let $\hat{\pi}(t)$ be the discontinuous variable change corresponding to $\hat{\vartheta}$ and $\hat{\theta}(s)$ be the inverse function. Then, the process $(\tilde{x}, \hat{\theta}, \tilde{u}, \hat{v}(s), \hat{\alpha}, 1)$, corresponding to the optimal process $(\hat{x}, \hat{u}, \hat{\vartheta})$ in Problem (4.2) by virtue of the formulas provided above, is optimal in Problem (4.17). Now, the conventional Pontryagin maximum principle may be applied to this process; see [7, 8]. Then, there exist a number $\lambda \geq 0$ and absolutely continuous functions $\tilde{\psi}(s)$, $\tilde{\psi}_\chi(s)$ such that:

$$\lambda + |\tilde{\psi}(s)| + |\tilde{\psi}_\chi(s)| \neq 0 \ \forall s \in [0, 1], \tag{4.18}$$

$$\begin{aligned}\tilde{\psi}(s) = \tilde{\psi}(0) - \int_0^s \hat{\alpha}(\varsigma)\frac{\partial H}{\partial x}(\tilde{x}(\varsigma), \tilde{u}(\varsigma), \tilde{\psi}(\varsigma), \hat{\theta}(\varsigma))d\varsigma \\ - \int_0^s (\hat{c} - \hat{\alpha}(\varsigma))\frac{\partial}{\partial x}\Big\langle Q(\tilde{x}(\varsigma), \tilde{\psi}(\varsigma), \hat{\theta}(\varsigma)), \hat{v}(\varsigma)\Big\rangle d\varsigma,\end{aligned} \tag{4.19}$$

$$\begin{aligned}\tilde{\psi}_\chi(s) = \tilde{\psi}_\chi(0) - \int_0^s \hat{\alpha}(\varsigma)\frac{\partial H}{\partial t}(\tilde{x}(\varsigma), \tilde{u}(\varsigma), \tilde{\psi}(\varsigma), \hat{\theta}(\varsigma))d\varsigma \\ - \int_0^s (\hat{c} - \hat{\alpha}(\varsigma))\frac{\partial}{\partial t}\Big\langle Q(\tilde{x}(\varsigma), \tilde{\psi}(\varsigma), \hat{\theta}(\varsigma)), \hat{v}(\varsigma)\Big\rangle d\varsigma,\end{aligned} \tag{4.20}$$

$$(\tilde{\psi}(0), -\tilde{\psi}(1), \tilde{\psi}_\chi(0), -\tilde{\psi}_\chi(1)) \in \lambda\varphi'(\hat{p}) + N_S(\hat{p}), \tag{4.21}$$

$$\begin{aligned}&\max_{\alpha\in[0,\hat{c}]} \ \max_{v\in K:\,|v|=1} \ \max_{u\in U}\Big(\alpha H(\tilde{x}(s), u, \tilde{\psi}(s), \hat{\theta}(s)) + \alpha\tilde{\psi}_\chi(s) \\ &+ (\hat{c} - \alpha)\Big\langle Q(\tilde{x}(s), \tilde{\psi}(s), \hat{\theta}(s)), v\Big\rangle\Big) = \\ &= \hat{\alpha}(s)\Big(H(\tilde{x}(s), \tilde{u}(s), \tilde{\psi}(s), \hat{\theta}(s)) + \tilde{\psi}_\chi(s)\Big) \\ &+ (\hat{c} - \hat{\alpha}(s))\Big\langle Q(\tilde{x}(s), \tilde{\psi}(s), \hat{\theta}(s)), \hat{v}(s)\Big\rangle = 0, \quad \text{a.a. } s \in [0, 1].\end{aligned} \tag{4.22}$$

Since $\hat{\alpha} \neq 0$, the nontriviality condition (4.18) is transformed into

$$\lambda + |\tilde{\psi}(s)| \neq 0 \ \ \forall s \in [0, 1]. \tag{4.23}$$

Optimal processes in Problems (4.2) and (4.17) are related by virtue of the formulas reproduced above. Let us demonstrate the relations between the corresponding Lagrange multipliers. Consider

$$\begin{cases}\psi(t) = \tilde{\psi}(\hat{\pi}(t)), \ \psi_\tau(s) = \tilde{\psi}(\hat{\gamma}_\tau(s)), \\ \phi(t) = -\tilde{\psi}_\chi(\hat{\pi}(t)), \ \phi_\tau(s) = -\tilde{\psi}_\chi(\hat{\gamma}_\tau(s)),\end{cases}$$

where $\hat{\gamma}_\tau(s)$ is the appropriate scaling function, which is defined similarly to the above. Then, obviously, by taking into account that $\tilde{x}(s) = \hat{x}(\hat{\theta}(s))$, $\tilde{u}(s) = \hat{u}(\hat{\theta}(s))$, the change of variable in Conditions (4.19), (4.21) and (4.23) yields conditions (4.11)–(4.14) directly. Condition (4.22) leads to both (4.15) and (4.16) after a simple variables separation. Indeed, by considering in (4.22) values $\alpha = \hat{c}$ and $\alpha = 0$ we obtain

$$\sup_{u \in U} H(\tilde{x}(s), u, \tilde{\psi}(s), \hat{\theta}(s)) + \tilde{\psi}_\chi(s) \leq 0,$$
$$\max_{v \in K:\, |v|=1} \left\langle Q(\tilde{x}(s), \tilde{\psi}(s), \hat{\theta}(s)), v \right\rangle \leq 0.$$

These inequalities are true for all $s \in [0, 1]$. From these, and (4.22),

$$\hat{\alpha}(s)\Big(H(\tilde{x}(s), \tilde{u}(s), \tilde{\psi}(s), \hat{\theta}(s)) + \tilde{\psi}_\chi(s)\Big) =$$

$$(\hat{c} - \hat{\alpha}(s))\Big\langle Q(\tilde{x}(s), \tilde{\psi}(s), \hat{\theta}(s)), \hat{v}(s)\Big\rangle = 0 \quad \text{a.a. } s \in [0, 1].$$

From here, and in view of (4.20), Conditions (4.15) and (4.16) follow straightforwardly in view of the variable change and the formulas provided in the first part of the proof which relate the processes of Problems (4.2) and (4.17).

The proof is complete. □

Remark 4.2 Surprisingly, in spite of the fact that the conventional part f of the dynamics does not affect the trajectory jump evolution, we can extract from the proof some additional information about the values of H at the points of discontinuity:

$$\sup_{u \in U} H(\hat{x}_\tau(s), u, \psi_\tau(s), \tau) \leq \phi_\tau(s) \;\; \forall\, s \in [0, 1] \;\forall\, \tau \in \mathrm{Ds}(|\hat{\vartheta}|).$$

Remark 4.3 From the maximum condition (4.14), we extract the time transversality conditions:

$$\phi(\hat{t}_0^+) = \sup_{u \in U} H(\hat{x}(\hat{t}_0^+), u, \psi(\hat{t}_0^+), \hat{t}_0),$$

$$\phi(\hat{t}_1^-) = \sup_{u \in U} H(\hat{x}(\hat{t}_1^-), u, \psi(\hat{t}_1^-), \hat{t}_1),$$

which coincide with the conventional time transversality conditions when impulsive part G is absent.

4.6 Exercises

Exercise 4.1 Let closed convex cone K be pointed. Show that for any impulsive control $(\mu; \{v_\tau\})$

$$v_\tau \neq 0 \quad \Leftrightarrow \quad \tau \in \mathrm{Ds}(|\mu|).$$

Exercise 4.2* Consider on $[0, 1]$ the ordinary differential equation

$$\dot{x} = G(x)v, \quad x(0) = x_0,$$

where $v : [0, 1] \to \mathbb{R}^k$ is some measurable function. The solution to this equation (assume that it exists) is denoted by $x(\cdot; v)$. Suppose that the Frobenius condition (3.5) is satisfied, that is, vector fields G^i, $i = 1, .., k$, pairwise commute. Let S be the simplex in $\mathbb{R}^k$, and let v_1, v_2 be any measurable functions with values in S such that

$$\int_0^1 v_1(s)ds = \int_0^1 v_2(s)ds.$$

Prove that $x(1; v_1) = x(1; v_2)$. (Use the material of Chap. 3.)

Exercise 4.3 Let $f(\cdot) : \mathbb{R} \to \mathbb{R}$ be a measurable function and μ be a nonnegative Borel measure. Prove that, by changing values of f on some subset of zero ℓ-measure, it is possible to make f Borel. Ensure then that f is $(\ell + \mu)$-measurable.

Exercise 4.4 Let $f \in \mathrm{L}_1([0, 1]; \mathbb{R})$ be a measurable integrable function and $\{\theta_i\}$ be a sequence of absolutely continuous functions such that $\theta_i \rightrightarrows \theta \in \mathrm{C}([0, 1]; \mathbb{R})$, where θ is absolutely continuous as well. Then, $f \circ \theta_i \to f \circ \theta$ in L_1.

Exercise 4.5 Let x satisfy on $[0, 1]$ the following ordinary differential equation

$$\dot{x} = G(x)v, \quad x(0) = x_0,$$

where $v : [0, 1] \to \mathbb{R}^k$ is a measurable bounded function such that

$$\int_0^1 v(s)ds = a \in \mathbb{R}^k.$$

By considering the discontinuous time variable change given by the inverse map

$$\theta_1(s) := \int_0^s \chi_D(\sigma)d\sigma,$$

where $D = \{s \in [0, 1] : v(s) \neq 0\}$, and χ_D is the characteristic function of a set, show that it is not restrictive to remove the set of points where $v(\cdot)$ vanishes. By subsequently applying the normalization time change, that is

$$\theta_2(s) = c \cdot \int_0^s |v(\sigma)|d\sigma,$$

where c is some positive number, find that there exist on $[0, 1]$ a control $\tilde{v}(\cdot)$, a number $\gamma \geq |a|$, and a trajectory $y(\cdot)$ such that $\dot{y} = G(y)\tilde{v}$, $y(0) = x_0$, the integral of $\tilde{v}$ equals again a, and $|\tilde{v}(s)| = \gamma$ for a.a. $s \in [0, 1]$. Furthermore, due to a certain absolutely continuous increasing function θ, show that one has $x \circ \theta = y$. Make use of Exercise 4.3 in the process.

Exercise 4.6 Prove Lemma 4.2.

Exercise 4.7 Construct the extension for (4.1). Prove the ambiguity of the arc selection. That is, prove that the assertion of Exercise 4.2 is not valid for system (4.1).

Exercise 4.8 Derive the Helly theorem from the Arzela–Ascoli theorem by using the Lebesgue discontinuous time variable change.

Exercise 4.9 For various practical applications, it may sometimes be important to minimize $\|\vartheta\|$ in (4.2) instead of $\varphi(p)$. Show that the new formulation is already embedded in Problem statement (4.2). Use the reduction of Stage 2 in the proof of Lemma 4.1.

Exercise 4.10 Show that set $\mathrm{Cs}(|\vartheta|)$ in (4.5) cannot be replaced with $\mathrm{Cs}(\mu)$.

References

1. Arutyunov, A., Karamzin, D.: Necessary conditions for a minimum in optimal impulse control problems (in Russian). Nonlinear dynamics and control collection of articles edited by S.V. Yemelyanov and S.K. Korovin. Moscow, Fizmatlit **4**(5), 205–240 (2004)
2. Arutyunov, A., Karamzin, D., Pereira, F.: On a generalization of the impulsive control concept: controlling system jumps. Discret. Contin. Dyn. Syst. **29**(2), 403–415 (2011)
3. Dykhta, V., Samsonyuk, O.: Optimal Impulse Control with Applications. Fizmatlit, Moscow (2000). [in Russian]
4. Fikhtengolts, G.: Course of Differential and Integral Calculus, vol. 3. Nauka, Moscow (1966). [in Russian]
5. Filippov, A.: On certain problems of optimal regulation. Bull. Mosc State Univ. Ser. Math. Mech. (2), 25–38 (1959)
6. Karamzin, D.: Necessary conditions of the minimum in an impulse optimal control problem. J. Math. Sci. **139**(6), 7087–7150 (2006)
7. Mordukhovich, B.: Variational analysis and generalized differentiation II. Applications. Grundlehren der Mathematischen Wissenschaften (Fundamental Principles of Mathematical Sciences), vol. 331. Springer, Berlin (2006)
8. Pontryagin, L., Boltyanskii, V., Gamkrelidze, R., Mishchenko, E.: Mathematical Theory of Optimal Processes, 1st edn. Transl. from the Russian ed. by L.W. Neustadt. Interscience Publishers, Wiley (1962)
9. Rishel, R.: An extended pontryagin principle for control systems, whose control laws contains measures. J. SIAM. Ser. A. Control **3**(2), 191–205 (1965)
10. Vidale, M., Wolfe, H.: An operations-research study of sales response to advertising. Oper. Res. **5**(3), 370–381 (1957)
11. Warga, J.: Variational problems with unbounded controls. J. SIAM. Ser. A. Control **3**(2), 424–438 (1965)

Chapter 5
Impulsive Control Problems with State Constraints

Abstract In this chapter, in the context of the impulsive extension of the optimal control problem, the state constraints are studied. That is, it is assumed that a certain closed subset of the state space is given while feasible arcs are not permitted to take values outside of it. This set is defined functionally in our considerations. It should be noted that the state constraints are in great demand in various engineering applications. For example, an iRobot cleaning a house should be able to avoid obstacles or objects that arise in its path. These obstacles are nothing but state constraints, while the task of avoiding the obstacle represents an important class of problems with state constraints. Evidently, there is a host of other engineering problems, in which the state constraints play an important role. The chapter deals with the same problem formulation as in the previous chapter; however, the state constraints of the above type are added. For this impulsive control problem, the Gamkrelidze-like maximum principle is obtained. Conditions for nondegeneracy of the maximum principle are presented. The chapter ends with eight exercises.

5.1 Introduction

In this chapter, as in the previous chapter, the focus is on the impulsive control problem resulting from the extension of (5), but, under some extra constraints, the so-called inequality state constraints. State constraints are defined by a certain smooth vector-valued function l which depends on state variable x and on time variable t, while the set of points

$$x \in \mathbb{R}^n \text{ satisfying } l(x,t) \le 0$$

is called state domain (note that it depends on time).

State constraints of this type are in high demand for a variety of engineering applications. For example, an iRobot, performing housecleaning, should be able to avoid collisions with obstacles or objects that appear in the way. These obstacles are mathematically modeled as state constraints, and therefore, the problem of avoiding

© Springer Nature Switzerland AG 2019

A. Arutyunov et al., *Optimal Impulsive Control*, Lecture Notes in Control and Information Sciences 477, https://doi.org/10.1007/978-3-030-02260-0_5

an obstacle is an important type of problem with state constraints. There are many other engineering applications where state constraints play an important role.

State constraints in the impulsive control context have been studied by different authors; see, e.g., [1–4]. In this chapter, another form of the Pontryagin maximum principle for impulsive control problems with state constraints is proposed. This form is based on the approach suggested in [5]. Despite a direct relation and a variable change, establishing the equivalence between the two known sets of the necessary optimality conditions for state-constrained problems, that is, the Gamkrelidze form, [5], (for its various developments, see also [6–11]) and the Dubovitskii–Milyutin form obtained by Dubovitskii and Milyutin in [12, 13] (for its various developments, see also [7, 8, 14–18], and for nonsmooth versions, see [19–21]), they are distinct, as the Gamkrelidze-like optimality conditions involve the extended Hamilton–Pontryagin function. The extended Hamilton–Pontryagin function, in comparison with the classical Hamilton–Pontryagin function, has an extra term given by the derivative of the state constraint function w.r.t. the control differential system. Therefore, this method requires the existence of the second-order derivatives for the state constraint function w.r.t. the state variables.

Despite this extra assumption imposed on the data of problem, the consideration of extended Hamilton–Pontryagin function $\bar{H}$ has some advantages with regard to application. A well-known example is the geodesic equation, see, e.g., [9, 22, 23], which immediately follows from the Gamkrelidze set of conditions. The Gamkrelidze approach also appears to be convenient also for the study of the continuity of the measured Lagrange multiplier under various regularity assumptions; see [10, 24–26]. Such a property is relevant to applications in light of indirect computational approach. Another useful outcome follows from the study of conditions for nondegeneracy and the conservation law in the presence of state constraints [27, 28]. The extended maximum condition relates the adjoint function and the measured multiplier: It allows us to express the measured multiplier via the adjoint function under a natural regularity condition.

The proof of main result presented in this chapter uses a different technique than that in Chap. 4. It consists of a rather direct approach, based on a certain penalization technique and on the Ekeland variational principle [29]. Due to this penalization method, it is possible to treat more general control problems with the right-hand part merely measurably w.r.t. the time variable. However, the method of discontinuous variable change is still utilized as an essential part of the proof.

Before proceeding to the problem statement, let us make an important note regarding the cone K. In Chap. 4, an arbitrary closed convex cone K was considered. However, it is not restrictive to consider the case in which the cone K in (5) lies entirely in first orthant $\mathbb{R}^k_+ := \{\xi \in \mathbb{R}^k : \xi^j \geq 0\}$. Indeed, it has been justified in the proof of Lemma 4.1 of Chap. 4 (see reduction (4.9) therein). Let us reproduce these arguments in the present context. Along with (5), which is under extension, consider the system

$$\dot{x} = f(x, u, t) + G(x, t)P\bar{v}, \quad \bar{v} \in \bar{K}. \tag{5.1}$$

Here, $P : \mathbb{R}^{2k} \to \mathbb{R}^k$ is a linear operator defined by $(P\xi)^j = \xi^{2j} - \xi^{2j-1}$, $j = 1, ..., k$, where $\xi = (\xi^1, \xi^2, ..., \xi^{2k})$, and

$$\bar{K} = \{\bar{v} \in \mathbb{R}^{2k}_+ : P\bar{v} \in K\}$$

is a closed and convex cone.

Note that any scalar function $r(t)$ can be represented as a difference of two nonnegative functions, i.e., $r = r^+ - r^-$, where $r^+(t) = \max\{r(t), 0\}$ and $r^-(t) = -\min\{r(t), 0\}$. For a given function $\bar{v}$, define $v := P\bar{v}$. Conversely, for a given function v, construct $\bar{v}$ by the formula $\bar{v}^{2j-1} := v^{j-}, \bar{v}^{2j} := v^{j+}$, $j = 1, ..., k$. This establishes the equivalence of (5) and (5.1) in the sense that the sets of admissible trajectories for (5) and (5.1) coincide. At the same time, this equivalence is ambiguous, since one control v may correspond to various controls $\bar{v}$. Nevertheless, all such $\bar{v}$ generate the same trajectory.

Due to these facts, in this chapter, the cone K is considered to lie solely in the first orthant. On the one hand, this somewhat simplifies notations (the measure ν and the set $V(\mu)$ in Chap. 4 disappear). On the other hand, in the unlikely event that the above-mentioned ambiguity is somehow crucial for a given application, one could easily transfer and apply the scheme of Chap. 4 to the considerations of the present chapter.

The exercises in this chapter are mostly technical. However, they will certainly be useful for those wishing to understand the details of the main proof. The proof of the main result itself is written in a somewhat shortened form in order to facilitate the understanding of the basic ideas, while some technical details are left as exercises.

5.2 Problem Statement

Consider the following optimal impulsive control problem with state constraints:

$$\begin{aligned}
\text{Minimize } \ & \varphi(p) \\
\text{subject to: } \ & dx = f(x, u, t)dt + G(x, t)d\vartheta, \\
& dy = d|\vartheta|, \ y(t_0) = 0, \\
& l(x, t) \leq 0, \\
& u(t) \in U, \ \text{range}(\vartheta) \subset K, \\
& p \in S, \ t \in T.
\end{aligned} \tag{5.2}$$

Here, the functions $\varphi : \mathbb{R}^{2n+1} \to \mathbb{R}^1$, $f : \mathbb{R}^n \times \mathbb{R}^m \times \mathbb{R}^1 \to \mathbb{R}^n$, $G : \mathbb{R}^n \times \mathbb{R}^1 \to \mathbb{R}^{n\times k}$, $l : \mathbb{R}^n \times \mathbb{R}^1 \to \mathbb{R}^d$ satisfy a certain smoothness assumption which is stated later, $T = [t_0, t_1]$ is a fixed time interval, $p = (x_0, x_1, y_1)$, with $x_0 = x(t_0)$, $x_1 = x(t_1)$, $y_1 = y(t_1)$ is the so-called endpoint vector, S is a closed set in $\mathbb{R}^{2n+1}$, U is a closed set in $\mathbb{R}^m$, K is a closed and convex cone embedded in first orthant $\mathbb{R}^k_+$, $\varphi(p)$ is the cost function to be minimized, the vector-valued function l defines the inequality state constraints, and $\vartheta = (\mu; \{v_\tau\})$ is the impulsive control. The measure $|\vartheta|$ designates the variation measure of the impulsive control, whereas y_1 becomes its total variation.

The notion of impulsive control has been defined in Chap. 4. However, the fact that K is embedded in $\mathbb{R}^k_+$ allows for simplification of definitions. Let us reproduce the notion of impulsive control by taking into account this extra information about cone K. The impulsive control consists of two components. While the first component, μ, is a vector-valued Borel measure with range in K (this means that $\mu(B) \in K$ for any Borel set $B \subseteq T$), the second component, $\{v_\tau\}$, is a family of measurable vector-valued functions defined on the interval $[0, 1]$, with values in K, and depending on the real parameter $\tau \in T$. Reproduce the properties of this family as well as the definition of trajectory in (5.2).

Consider a Borel vector-valued measure μ such that range$(\mu) \subset K$. Take a number $\tau \in T$. Denote by $W_\tau(\mu) \subset L_\infty([0, 1]; \mathbb{R}^k)$ the set of functions $v : [0, 1] \to K$ satisfying the following two conditions:

(i) $\sum\limits_{j=1}^{k} v^j(s) = \sum\limits_{j=1}^{k} \mu^j(\tau)$, a.a. $s \in [0, 1]$;

(ii) $\int\limits_0^1 v^j(s)ds = \mu^j(\tau)$, $j = 1, \ldots, k$.

Here, $\mu^j(\tau) = \mu^j(\{\tau\})$ is the value of μ^j at the single point set $\{\tau\}$, and the symbol a.a. s means "almost all s." Note that $W_\tau(\mu) = \{0\}$ whenever $\mu^j(\tau) = 0, \forall j$.

Definition 5.1 The pair $\vartheta = (\mu; \{v_\tau\})$ is said to be impulsive control, provided $v_\tau \in W_\tau(\mu)\,\forall \tau \in T$. The family $\{v_\tau\}$ is termed attached to the vector-valued measure μ.

Definition 5.2 The variation $|\vartheta|$ of the impulsive control $\vartheta = (\mu; \{v_\tau\})$ is said to be the variation measure $|\mu| := \sum_{j=1}^{k} \mu^j$.

So, the impulsive control is the measure complemented with the attached family. The attached family is defined by the above relations (i) and (ii). As usual, the function $u(\cdot)$ in (5.2) stands for the conventional control. It is measurable and essentially bounded w.r.t. Lebesgue measure ℓ. Let us recall the concept of trajectory.

Take a point $\tau \in T$, a function $v \in L_\infty([0, 1]; \mathbb{R}^k)$, and an arbitrary vector $a \in \mathbb{R}^n$. Denote by $z(\cdot) := z(\cdot; \tau, a, v)$ the solution to the following dynamical system

$$\begin{cases} \dot{z}(s) = G(z(s), \tau)v(s), & s \in [0, 1], \\ z(0) = a. \end{cases}$$

The function of bounded variation $x(\cdot)$ on the interval T is termed solution to the differential equation in (5.2), corresponding to the control (u, ϑ), where $\vartheta = (\mu; \{v_\tau\})$, and the initial point x_0, provided that $x(t_0) = x_0$ and, for every $t \in (t_0, t_1]$,

$$x(t) = x_0 + \int_{t_0}^{t} f(x, u, \varsigma)d\varsigma + \int_{[t_0,t]} G(x, \varsigma)d\mu_c + \sum_{\tau \in \text{Ds}(\mu):\, \tau \le t} \Big(x_\tau(1) - x(\tau^-)\Big), \tag{5.3}$$

where $x_\tau(\cdot) := z(\cdot; \tau, x(\tau^-), v_\tau)$, and the measure μ_c signifies the continuous component of μ. Note that the sum in (5.3) is well defined as the atomic set of a Borel measure is countable. Moreover, the summation is absolutely convergent due to the fact that the sum of total distances jumped is bounded.

The state constraints $l(x, t) \le 0$ in (5.2) should be understood in a broader sense than that of a conventional inequality. This is due to the presence of the impulsive control. Let us define the precise meaning of this inequality. Consider an impulsive control $\vartheta = (\mu; \{v_\tau\})$ and an admissible trajectory $x(\cdot)$ corresponding to ϑ.

The attached functions $x_\tau(\cdot)$, introduced above, are parts of the extended trajectory defining the paths in the state space joining the endpoints of each jump. Therefore, they should also be subject to the state constraint just as the main trajectory $x(\cdot)$. Then, the inequality $l(x, t) \le 0$ is enriched with the following generalized sense:

$$l(x, t) \le 0 \quad \Leftrightarrow \quad \begin{cases} l(x(t), t) \le 0, & \text{a.a. } t \in T, \\ l(x_\tau(s), \tau) \le 0, & \text{a.a. } s \in [0, 1], \ \forall \tau \in \text{Ds}(\mu). \end{cases}$$

The notation $dy = d|\vartheta|$, according to what has been stated, stands for $y(\cdot) = F(\cdot; |\mu|)$. The extra state variable y is, on the one hand, redundant in the formulation because it can always be expressed through the basic part of dynamics given for variable x. On the other hand, the explicit availability of the constrained total variation of impulsive control in the problem statement, as has been noted in the previous chapters, is useful with regard to applications. Moreover, the question of the new type of condition in the maximum principle generated by the constraints imposed on the total variation is interesting in itself. For this reason, the variable y is included in the problem formulation.

The quadruple (x, y, u, ϑ) is said to be a control process, provided that $dy = d|\vartheta|$ and (5.3) holds. A control process is said to be admissible if all the constraints of problem (5.2) are satisfied. An admissible process $(\hat{x}, \hat{y}, \hat{u}, \hat{\vartheta})$ is said to be optimal provided that, for any admissible process (x, y, u, ϑ), the inequality $\varphi(\hat{p}) \le \varphi(p)$ holds, where $\hat{p} = (\hat{x}(t_0), \hat{x}(t_1), \hat{y}(t_1))$.

From henceforth, the following hypotheses are assumed:

(H1) The vector-valued function f is differentiable w.r.t. x for all u, and for a.a. t. Function f and its partial derivative w.r.t. x are Lebesgue measurable w.r.t. t for all x, u, and, on any bounded set, continuous w.r.t. x, u uniformly w.r.t. x, u, t, and bounded. The matrix-valued function G is differentiable w.r.t. x for all t. Function G and its partial derivative w.r.t. x are continuous w.r.t. x, t. The mapping φ is continuously differentiable. The mapping l is twice continuously differentiable.

(H2) The mappings f, G, and φ are continuously differentiable. The mapping l is twice continuously differentiable.

5.3 Maximum Principle in Gamkrelidze's Form

Denote by $\bar{H}$ the extended Hamilton–Pontryagin function

$$\bar{H}(x, u, \psi, \eta, t) := \langle \psi, f(x, u, t)\rangle - \langle \eta, r(x, u, t)\rangle,$$

where $r(x, u, t) := \frac{\partial l}{\partial x}(x, t) f(x, u, t) + \frac{\partial l}{\partial t}(x, t)$. Denote by $\bar{Q}$ the vector-valued function

$$\bar{Q}(x, \psi, \eta, t) := G^*(x, t)\psi - M^*(x, t)\eta,$$

where

$$M(x, t) := \frac{\partial l}{\partial x}(x, t) G(x, t)$$

is matrix-valued.

Theorem 5.1 *Let $(\hat{x}, \hat{y}, \hat{u}, \hat{\vartheta})$, where $\hat{\vartheta} = (\hat{\mu}; \{\hat{v}_\tau\})$, be an optimal process in Problem (5.2). Assume that Hypothesis (H1) is in force.*

Then, there exist a number $\lambda \geq 0$, a number $\omega \in \mathbb{R}^1$, a vector-valued function of bounded variation ψ, and a vector-valued function η, both defined on T, and for every point $\tau \in \mathrm{Ds}(\hat{\mu})$, there exist an absolutely continuous vector-valued function ψ_τ and a vector-valued function η_τ, both defined on $[0, 1]$, such that,

$$\lambda + |\omega| + |\psi(t_0)| + |\eta(t_0)| > 0, \tag{5.4}$$

$$\begin{aligned} \psi(t) = \psi(t_0) - \int_{t_0}^{t} \frac{\partial \bar{H}}{\partial x}(\varsigma) d\varsigma - \int_{[t_0,t]} \frac{\partial}{\partial x}\left\langle \bar{Q}(\varsigma), d\hat{\mu}_c \right\rangle \\ + \sum_{\tau \in \mathrm{Ds}(\hat{\mu}):\, \tau \leq t} \left(\psi_\tau(1) - \psi(\tau^-)\right), \quad \forall t \in (t_0, t_1], \end{aligned} \tag{5.5}$$

$$\begin{cases} \dot{\psi}_\tau(s) = -\frac{\partial}{\partial x}\langle \bar{Q}_\tau(s), \hat{v}_\tau(s)\rangle, \\ \psi_\tau(0) = \psi(\tau^-), \; s \in [0, 1], \; \forall \tau \in \mathrm{Ds}(\hat{\mu}), \end{cases} \tag{5.6}$$

$$\left(\psi(t_0) - \frac{\partial l}{\partial x}(t_0)\eta(t_0), -\psi(t_1), \omega\right) \in \lambda \varphi'(\hat{p}) + N_S(\hat{p}), \tag{5.7}$$

$$\max_{u \in U} \left\{\bar{H}(u, t)\right\} = \bar{H}(t), \quad a.a.\, t \in T, \tag{5.8}$$

$$\begin{cases} \max\limits_{v \in K \cap S_{\mathbb{R}^k}} \langle \bar{Q}(t), v\rangle \leq \omega & \forall t \in T, \\ \max\limits_{v \in K \cap \hat{v}_\tau S_{\mathbb{R}^k}} \langle \bar{Q}_\tau(s), v\rangle = \langle \bar{Q}_\tau(s), \hat{v}_\tau(s)\rangle = \hat{v}_\tau \omega \; a.a.\, s \in [0, 1], & \\ \int_A \langle \bar{Q}(t), d\hat{\mu}_c\rangle = \omega \cdot |\hat{\mu}_c|(A) & \text{for all Borel sets } A, \end{cases} \tag{5.9}$$

where $\hat{v}_\tau = |\hat{\mu}|(\tau)$, $\forall \tau \in \mathrm{Ds}(\hat{\mu})$.

The vector-valued function $\eta(t) = (\eta^1(t), \ldots, \eta^d(t))$ *satisfies the following conditions:*

(a) *Each one of the functions* η^j *is constant on any time interval* $\Sigma = [\sigma_1, \sigma_2] \subseteq T$*, on which the optimal trajectory* $\hat{x}(t)$ *lies in the interior of the set defined by* j*th state constraint, that is, when* $l^j(s) < 0$, $\forall s \in \Sigma$.
(b) *The functions* η^j *are left continuous on the interval* (t_0, t_1).
(c) *The functions* η^j *are decreasing and* $\eta^j(t_1) = 0$.

The vector-valued functions $\eta_\tau(s) = (\eta_\tau^1(s), \ldots, \eta_\tau^d(s))$, $\tau \in \mathrm{Ds}(\hat{\mu})$*, satisfy similar conditions:*

(a′) *Each one of the functions* η_τ^j *is constant on any time interval* $\Sigma \subseteq [0, 1]$*, on which the optimal trajectory* $\hat{x}_\tau(s)$ *lies in the interior of the set defined by* j*th state constraint, that is, when* $l_\tau^j(s) < 0$, $\forall s \in \Sigma$.
(b′) *The functions* η_τ^j *are left continuous on the interval* $(0, 1)$.
(c′) *The functions* η_τ^j *are decreasing, and* $\eta_\tau(0) = \eta(\tau^-)$ *and* $\eta_\tau(1) = \eta(\tau^+)$.

Here, as usual, if any of the arguments x, u, ψ, and η of a function is omitted, then it means that the extremal (w.r.t. the maximum principle) values $\hat{x}(t)$, $\hat{u}(t)$, $\psi(t)$, or $\eta(t)$, respectively, replace the omitted arguments. For example,

$$\bar{H}(t) = \bar{H}(\hat{x}(t), \hat{u}(t), \psi(t), \eta(t), t), \quad \bar{H}(u, t) = \bar{H}(\hat{x}(t), u, \psi(t), \eta(t), t).$$

Similarly, the subscript τ of a function means that the extremal values $\hat{x}_\tau(s)$, $\psi_\tau(s)$, or $\eta_\tau(s)$ attached to the discontinuity point τ substitute the omitted arguments. For example, $\bar{Q}_\tau(s) = \bar{Q}(\hat{x}_\tau(s), \psi_\tau(s), \eta_\tau(s), \tau)$. The same notation is used for the partial derivatives of functions w.r.t. x.

Proof. Let the process $(\hat{x}, \hat{y}, \hat{u}, \hat{\vartheta})$ be optimal in Problem (5.2). It is not restrictive to assume that $\varphi(\hat{p}) = 0$ and also that $\hat{u}$, f, and f_x' are $(\ell + |\hat{\vartheta}|)$-measurable w.r.t. t for all x and u. This may always be achieved by changing the values of the functions on a zero ℓ-measure subset. In order to simplify the presentation, assume that $d = 1$, that is, the case of a scalar-valued function $l(x, t)$. The case $d > 1$ can be considered by virtue of the same arguments as below.

Consider the discontinuous time variable change $\hat{\pi}(\cdot) : T \to [0, 1]$ related to the optimal impulsive control $\hat{\vartheta}$:

$$\hat{\pi}(t) = \frac{t - t_0 + |\hat{\vartheta}|([t_0, t])}{c} \quad \text{for } t > t_0, \quad \hat{\pi}(t_0) = 0,$$

where $c = t_1 - t_0 + |\hat{\vartheta}|(T)$. The inverse to $\hat{\pi}$ function is denoted by $\hat{\theta}(s) : [0, 1] \to T$.

Consider

$$\hat{m}_1(t) = \frac{c \cdot d\ell}{d(\ell + |\hat{\vartheta}|)}, \quad \text{and } \hat{m}_2(t) = \frac{c \cdot d\hat{\mu}_c}{d(\ell + |\hat{\vartheta}|)},$$

to be the Radon–Nikodym derivatives, respectively, of the measures ℓ and $\hat{\mu}_c$ w.r.t. Lebesgue–Stieltjes measure $\ell + |\hat{\vartheta}|$ multiplied by the number c. Both functions are $(\ell + |\hat{\vartheta}|)$-measurable and essentially bounded.

Let $\hat{\Gamma}_\tau = [\hat{\pi}(\tau^-), \hat{\pi}(\tau^+)]$, $\tau \in \mathrm{Ds}(\hat{\mu})$, and define

$$\hat{\alpha}(\varsigma) := \begin{cases} \hat{m}_1(\hat{\theta}(\varsigma)), & \text{if } \varsigma \notin \bigcup\limits_{\tau \in \mathrm{Ds}(\hat{\vartheta})} \hat{\Gamma}_\tau, \\ 0, & \text{otherwise}, \end{cases}$$

$$\hat{\beta}(\varsigma) := \begin{cases} c \cdot \ell^{-1}(\hat{\Gamma}_\tau)\hat{v}_\tau(\hat{\xi}_\tau(\varsigma)), & \text{if } \exists\, \tau \in \mathrm{Ds}(\hat{\mu}) \text{ s.t. } \varsigma \in \hat{\Gamma}_\tau, \\ \hat{m}_2(\hat{\theta}(\varsigma)), & \text{otherwise}, \end{cases}$$

where $\hat{\xi}_\tau : \hat{\Gamma}_\tau \to [0, 1]$ is given by $\hat{\xi}_\tau(s) = \frac{s - \hat{\pi}(\tau^-)}{\ell(\hat{\Gamma}_\tau)}$.

Note that (5.4)–(5.9) are equivalent to the following conditions:

$$\lambda + |\omega| + |\psi^{\mathrm{ext}}(0)| + \eta^{\mathrm{ext}}(0) > 0, \tag{5.10}$$

$$\psi^{\mathrm{ext}}(s) = \psi^{\mathrm{ext}}(0) - \int_0^s \frac{\partial \bar{H}^{\mathrm{ext}}}{\partial x}(\varsigma)\hat{\alpha}(\varsigma)d\varsigma - \int_0^s \frac{\partial}{\partial x}\langle \bar{Q}^{\mathrm{ext}}(\varsigma), \hat{\beta}(\varsigma)\rangle d\varsigma, \quad \forall s, \tag{5.11}$$

$$\left(\psi^{\mathrm{ext}}(0) - \frac{\partial l^{\mathrm{ext}}}{\partial x}(0)\eta^{\mathrm{ext}}(0), -\psi^{\mathrm{ext}}(1), \omega\right) \in \lambda\varphi'(\hat{p}) + N_S(\hat{p}), \tag{5.12}$$

$$\max_{u \in U}\left\{\hat{\alpha}(s)\bar{H}^{\mathrm{ext}}(u, s)\right\} = \hat{\alpha}(s)\bar{H}^{\mathrm{ext}}(s), \quad \text{a.a. } s, \tag{5.13}$$

$$\max_{v \in K}\left\{\langle \bar{Q}^{\mathrm{ext}}(s), v\rangle - \omega|v|\right\} = \langle \bar{Q}^{\mathrm{ext}}(s), \hat{\beta}(s)\rangle - \omega|\hat{\beta}(s)| = 0, \quad \text{a.a. } s, \tag{5.14}$$

where $s \in [0, 1]$, and whenever the function ϕ of x, u, ψ, η, and t has the superscript "ext", the extended extremal values replace the omitted arguments, that is,

$$\phi^{\mathrm{ext}}(s) = \begin{cases} \phi(\hat{x}_\tau(\hat{\xi}_\tau(s)), \hat{u}(\hat{\theta}(s)), \psi_\tau(\hat{\xi}_\tau(s)), \eta_\tau(\hat{\xi}_\tau(s)), \hat{\theta}(s)), & \text{if } s \in \hat{\Gamma}_\tau \text{ and } \tau \in \mathrm{Ds}(\hat{\mu}), \\ \phi(\hat{x}(\hat{\theta}(s)), \hat{u}(\hat{\theta}(s)), \psi(\hat{\theta}(s)), \eta(\hat{\theta}(s)), \hat{\theta}(s)), & \text{otherwise}. \end{cases}$$

This equivalence is valid in view of the discontinuous time variable change. The proof is straightforward. Thus, in order to prove the theorem, it is sufficient to derive Conditions (5.10)–(5.14) and establish the desired properties of the function η^{ext} analogous to those stated in (a)–(c).

Let us proceed with the details. Assume at the outset that the set U is bounded, i.e., $\exists\, \gamma > 0$ such that $U \subseteq \gamma B_{\mathbb{R}^m}$, where B_X stands for the unit ball in space X.

Note that Problem (5.2) is equivalent to the following problem:

$$
\begin{aligned}
\text{Minimize } & \varphi(p)\\
\text{subject to: } & dx = f(x,u,t)dt + G(x,t)d\vartheta,\\
& dy = d|\vartheta|,\\
& d\chi = r(x,u,t)dt + M(x,t)d\vartheta,\\
& \chi(t) \le 0,\\
& \chi_0 = l(x_0,t_0),\\
& p = (x_0,x_1,y_1) \in S,\\
& y(t_0) = 0,\ u(t) \in U.
\end{aligned}
\tag{5.15}
$$

This is so due to the formula

$$
l(x(t),t) = l(x_0,t_0) + \int_{t_0}^{t} r(x(\varsigma),u(\varsigma),\varsigma)d\varsigma + \int_{[t_0,t]} M(x(\varsigma),\varsigma)d\vartheta,
$$

which is simple to verify by performing the discontinuous time variable change. Thus, the control process $(\hat{x},\hat{y},\hat{u},\hat{\vartheta})$ is optimal in Problem (5.15).

It is also a straightforward task (see Exercise 5.1) to establish that there exists a sequence of nonimpulsive conventional controls $\bar{v}_i \in L_\infty(T;\mathbb{R}^k)$, such that $\bar{v}_i(t) \in K$, a.a. t, $\bar{v}_i \xrightarrow{w} |\hat{\vartheta}|$, $\bar{\zeta}_i^{\text{ext}}(s) \rightrightarrows \hat{\zeta}^{\text{ext}}(s)$, where $\frac{d\bar{v}_i}{dt} = |\bar{v}_i(t)|$, $\bar{\zeta}_i$ and $\hat{\zeta}$ are the solutions to the simplest system $d\zeta = d\vartheta$, $\zeta(t_0) = 0$, corresponding to $\bar{v}_i$, $\hat{\vartheta}$, and the extension is considered in the above sense w.r.t. $\bar{v}_i$ and $\hat{\vartheta}$, respectively. In other words, we have

$$
\bar{\zeta}_i(\bar{\theta}_i(s)) = \int_{t_0}^{\bar{\theta}_i(s)} \bar{v}_i(\varsigma)d\varsigma \rightrightarrows \hat{\zeta}^{\text{ext}}(s),
$$

where $\bar{\theta}_i(s)$ is the inverse function of

$$
\bar{\pi}_i(t) = \left(t - t_0 + \int_{t_0}^{t} |\bar{v}_i(\varsigma)|d\varsigma\right)\cdot\left(t_1 - t_0 + \int_{t_0}^{t_1} |\bar{v}_i(\varsigma)|d\varsigma\right)^{-1}.
$$

It is clear that $\bar{\theta}_i(s) \rightrightarrows \hat{\theta}(s)$. Then, trajectories $(\bar{x}_i(t),\bar{y}_i(t))$ corresponding to $(\hat{x}_0,\hat{u},\bar{v}_i)$ converge to $(\hat{x}(t),\hat{y}(t))$ as $i \to \infty$, albeit in the extended sense, i.e., $\bar{x}_i(\bar{\theta}_i(s)) \rightrightarrows \hat{x}^{\text{ext}}(s)$, and $\bar{y}_i(\bar{\theta}_i(s)) \rightrightarrows \hat{y}^{\text{ext}}(s)$.

Define $\kappa_i = i + \operatorname{ess\,sup}_{t\in T} |\bar{v}_i(t)|^2$. Denote by $\mathscr{M}$ the set of triples $(x_0,u(\cdot),v(\cdot))$ such that $u(t) \in U$, $v(t) \in K \cap \kappa_i B_{\mathbb{R}^k}$ a.a. $t \in T$, there exists on T trajectory $x(t)$ such that $x(t_0) = x_0$ and $\dot{x}(t) = f(x(t),u(t),t) + G(x(t),t)v(t)$, a.a. $t \in T$, and inequality $|x(t)| \le \|\hat{x}^{\text{ext}}\|_C + 1\ \forall\, t \in T$ is satisfied. The set $\mathscr{M} \subseteq \mathbb{R}^n \times L_1(T;\mathbb{R}^m) \times L_1(T;\mathbb{R}^k)$ is closed (Exercise 5.3), and thereby, it becomes the complete metric space if it is endowed with the metric generated by norm $|x_0| + \|(u,v)\|_{L_1}$.

Let $\varepsilon > 0$ be a positive number. Denote $\varphi_\varepsilon(p) := \left(\varphi(p) + \frac{\varepsilon^2}{2}\right)^+$, where $a^+ := \max\{a, 0\}$. Let α, β be nonnegative numbers. Define

$$\Delta(\alpha, \beta) := \begin{cases} \alpha\beta^{-3}, & \text{for } \beta > 0, \\ 1, & \text{for } \alpha \neq 0, \text{ and } \beta = 0, \\ 0, & \text{for } \alpha = \beta = 0. \end{cases}$$

Note that Δ is lower semicontinuous on $\mathbb{R}^2_+$ (Exercise 5.4).

Consider a functional on $\mathrm{L}_1(T; \mathbb{R}^m) \times \mathrm{L}_1(T; \mathbb{R}^k)$ defined as follows

$$D_i(v(\cdot)) := \int_{t_0}^{t_1} V_i(\zeta(t), v(t), t)dt,$$

where

$$V_i(\zeta, v, t) := |\zeta - \bar{\zeta}_i(t)|^2 \cdot \left(1 + \frac{|v| + |\bar{v}_i(t)|}{2}\right),$$

$$\zeta(t) = \int_{t_0}^{t} v(\varsigma)d\varsigma, \quad \bar{\zeta}_i(t) = \int_{t_0}^{t} \bar{v}_i(\varsigma)d\varsigma.$$

On space $\mathscr{M}$, define

$$\Phi_i(x_0, u(\cdot), v(\cdot), \varepsilon) := \varphi_\varepsilon(p) + \varepsilon D_i(v(\cdot)) + \Delta\left((\mathrm{dist}(p, S))^2, \varphi_\varepsilon(p)\right) + \Delta\left(\int_{t_0}^{t_1} (\chi^+(t))^2 \left(1 + \frac{|v| + |\bar{v}_i(t)|}{2}\right) dt, \varphi_\varepsilon(p)\right).$$

Here, $p = (x(t_0), x(t_1), y(t_1))$ and x, and χ are solutions to

$$\dot{x} = f(x, u(t), t) + G(x, t)v(t),$$
$$\dot{\chi} = r(x, u(t), t) + M(x, t)v(t),$$

on T, with $\chi(t_0) = l(x_0, t_0)$.

The functional Φ_i is nonnegative and lower semicontinuous on $\mathscr{M}$. This follows directly from the definition. Note that, for every fixed ε, we have $\Phi_i(\hat{x}_0, \hat{u}, \bar{v}_i, \varepsilon) \to \frac{\varepsilon^2}{2}$ as $i \to \infty$. Then, it is not restrictive to assume that $\Phi_i(\hat{x}_0, \hat{u}, \bar{v}_i, \varepsilon_i) \le \varepsilon_i^2$, where $\varepsilon_i := \frac{1}{i} \downarrow 0$.

Consider the following problem

$$\text{Minimize } \Phi_i(x_0, u(\cdot), v(\cdot), \varepsilon_i)$$
$$\text{subject to: } (x_0, u, v) \in \mathscr{M}.$$

Let us apply the Ekeland variational principle, [29], to this problem at the point $(\hat{x}_0, \hat{u}, \bar{v}_i)$. Then, for every i, there exists an element $(x_{0,i}, u_i, v_i) \in \mathcal{M}$ such that

$$\Phi_i(x_{0,i}, u_i, v_i, \varepsilon_i) \leq \Phi_i(\hat{x}_0, \hat{u}, \bar{v}_i, \varepsilon_i) \leq \varepsilon_i^2, \tag{5.16}$$

$$|x_{0,i} - \hat{x}_0| + \int_{t_0}^{t_1} |u_i(t) - \hat{u}(t)|dt + \int_{t_0}^{t_1} |v_i(t) - \bar{v}_i(t)|dt \leq \varepsilon_i, \tag{5.17}$$

and the triple $(x_{0,i}, u_i(\cdot), v_i(\cdot))$ is the unique solution to the following control problem

$$\begin{aligned}
\text{Minimize}\, & \varphi_{\varepsilon_i}(p) + z_0^{-3}(\text{dist}(p, S))^2 \\
& + \int_{t_0}^{t_1} \frac{(\chi^+(t))^2}{z^3}\Big(1 + \frac{|v| + |\bar{v}_i(t)|}{2}\Big)dt + \varepsilon_i|x_0 - x_{0,i}| \\
& + \varepsilon_i \int_{t_0}^{t_1} (|u - u_i(t)| + |v - v_i(t)| + V_i(\zeta, v, t))\, dt, \\
\text{subject to: } & \dot{x} = f(x, u, t) + G(x, t)v, \\
& \dot{y} = |v|, \quad y_0 = 0, \\
& \dot{z} = 0, \quad z_0 = \varphi_{\varepsilon_i}(p), \\
& \dot{\zeta} = v, \quad \zeta_0 = 0, \\
& \dot{\chi} = r(x, u, t) + M(x, t)v, \quad \chi_0 = l(x_0, t_0), \\
& z_0 > 0, \ |x(t)| \leq \|\hat{x}^{\text{ext}}\|_C + 1, \\
& u(t) \in U, \quad v(t) \in K \cap \kappa_i B_{\mathbb{R}^k}, \quad \text{a.a. } t \in T.
\end{aligned} \tag{5.18}$$

The optimal control process of (5.18) is denoted by $(x_i, y_i, z_i, \zeta_i, \chi_i, u_i, v_i)$.

Let us explain why it is assumed above that $\varphi_{\varepsilon_i}(p_i) > 0$, where $p_i = (x_{0,i}, x_{1,i}, y_{1,i})$, and, as a consequence, why z_0 is considered greater than zero. Indeed, if $\varphi_{\varepsilon_i}(p_i) = 0$, then either the state or the endpoint constraints are violated in Problem (5.2). Let the endpoint constraints be violated. Then,

$$\Delta\left((\text{dist}(p_i, S))^2, \varphi_{\varepsilon_i}(p_i)\right) = \Delta\left((\text{dist}(p_i, S))^2, 0\right) = 1,$$

but, in view of (5.16), this is impossible as $i > 1$. Therefore, $\varphi_{\varepsilon_i}(p_i) > 0$.

From (5.16), we deduce that

$$\int_{t_0}^{t_1} V_i(t)dt \to 0$$

as $i \to \infty$, where $V_i(t)$ stands for $V_i(\zeta_i(t), v_i(t), t)$. From (5.17), it follows $x_{0,i} \to \hat{x}_0$, $u_i \to \hat{u}$ in L_1. Then, Exercises 5.5, and 5.6 (see also Lemma 6.1 of Chap. 6 for more general assertions) yield that $x_i(\theta_i(s)) \rightrightarrows \hat{x}^{\text{ext}}(s)$, where θ_i is the inverse to the variable change

$$\pi_i(t) = \left(\int_{t_0}^{t_1} (1 + |v_i(\varsigma)|) d\varsigma \right)^{-1} \int_{t_0}^{t} (1 + |v_i(\varsigma)|) d\varsigma.$$

Therefore, for i sufficiently large, the state constraint in (5.18), that is inequality $|x(t)| \le \|\hat{x}^{\text{ext}}\|_C + 1$, is not active at any t. Moreover, Problem (5.18) is conventional as it does not comprise impulsive controls. Then, for every i, there exist a number $\lambda_i > 0$ and absolutely continuous conjugate functions ψ_i^x, ψ_i^y, ψ_i^z, ψ_i^ζ, and ψ_i^χ such that the maximum principle from [30], Theorem 6.27, is satisfied.

Consideration of conditions of this maximum principle suggests the following time variable change:

$$\pi_i^*(t) = \left(\int_{t_0}^{t_1} m_i(\varsigma) d\varsigma \right)^{-1} \int_{t_0}^{t} m_i(\varsigma) d\varsigma,$$

where

$$m_i(t) = 1 + \frac{|v_i(t)| + |\bar{v}_i(t)|}{2}.$$

By virtue of this variable change, the conditions of the maximum principle for Problem (5.18) become

$$\begin{cases} \psi_i^{\text{ext}}(s) = \psi_i^{\text{ext}}(0) - \displaystyle\int_0^s \frac{\partial \bar{H}_i^{\text{ext}}}{\partial x}(\varsigma)\alpha_i(\varsigma) d\varsigma - \int_0^s \frac{\partial}{\partial x} \langle \bar{Q}_i^{\text{ext}}(\varsigma), \beta_i(\varsigma) \rangle d\varsigma, \\ \xi_i(s) \quad = \displaystyle\int_s^1 3\lambda_i \frac{[\chi_i^+(\theta_i^*(\varsigma))]^2}{z_i^4} d\varsigma, \\ \sigma_i(s) \quad = \displaystyle\int_s^1 2\lambda_i \varepsilon_i \left(\zeta_i^{\text{ext}}(\varsigma) - \bar{\zeta}_i^{\text{ext}}(\varsigma) \right) \alpha_i(\varsigma) d\varsigma, \\ \eta_i^{\text{ext}}(s) \quad = \displaystyle\int_s^1 2\lambda_i \frac{\chi_i^+(\theta_i^*(\varsigma))}{z_i^3} d\varsigma, \end{cases} \tag{5.19}$$

$$\begin{aligned} (\psi_i^{\text{ext}}(0), -\psi_i^{\text{ext}}(1), \omega_i) & \\ \in \big(\lambda_i - \xi_i(0) - 3\rho_i z_{0,i}^{-1} \operatorname{dist}(p_i, S)\big) & \varphi'(p_i) + 2\rho_i \partial \operatorname{dist}(p_i, S) \\ & + \left(\frac{\partial l_i^{\text{ext}}}{\partial x}(0) \eta_i^{\text{ext}}(0), 0, 0 \right) + (\lambda_i \varepsilon_i B_{\mathbb{R}^n}, 0, 0), \end{aligned} \tag{5.20}$$

$$\begin{cases} \max_{v \in K \cap \kappa_i B_{\mathbb{R}^k}} \max_{u \in U} \alpha_i(s) \Big\{ \bar{H}_i^{\text{ext}}(u, s) + \langle \bar{Q}_i^{\text{ext}}(s), v \rangle - \omega_i |v| + \langle \sigma_i(s), v \rangle \\ \qquad -\lambda_i \varepsilon_i \left(|u - u_i^{\text{ext}}(s)| + |v - v_i^{\text{ext}}(s)| + V_i^{\text{ext}}(u, v, s) \right) \Big\} \\ = \alpha_i(s) \bar{H}_i^{\text{ext}}(s) + \langle \bar{Q}_i^{\text{ext}}(s), \beta_i(s) \rangle - \omega_i |\beta_i(s)| + \langle \sigma_i(s), \beta_i(s) \rangle \\ \qquad -\alpha_i(s) \lambda_i \varepsilon_i V_i^{\text{ext}}(s), \quad \text{a.a. } s \in [0, 1], \end{cases} \tag{5.21}$$

$$\lambda_i + \|\psi_i^{\text{ext}}\|_{\mathrm{C}} + \eta_i^{\text{ext}}(0) + |\omega_i| + \rho_i = 1. \tag{5.22}$$

Here, $\theta_i^* = \pi_i^{*-1}$, $\psi_i^{\text{ext}}(s) = \psi_i^x(\theta_i^*(s))$, $\omega_i = -\psi_i^y(\theta_i^*(s)) \equiv \text{const}$, $\xi_i(s) = \psi_i^z(\theta_i^*(s))$, $\sigma_i(s) = \psi_i^\zeta(\theta_i^*(s))$, $\eta_i^{\text{ext}}(s) = -\psi_i^\chi(\theta_i^*(s))$, $\rho_i = \lambda_i z_{0,i}^{-3} \operatorname{dist}(p_i, S)$, $\alpha_i(s) = \dfrac{a_i}{m_i(\theta_i^*(s))}$, $\beta_i(s) = \dfrac{a_i v_i(\theta_i^*(s))}{m_i(\theta_i^*(s))}$, and $a_i = t_1 - t_0 + \|m_i\|_{\mathrm{L}_1}$.

Here, whenever a function above has the subscript i and the superscript "ext", it means that the extremal values for Problem (5.18) replace the omitted arguments, in terms of the above-performed variable change π_i^*. For example,

$$\bar{H}_i^{\text{ext}}(u, s) = \bar{H}(x_i(\theta_i^*(s)), u, \psi_i^{\text{ext}}(s), \eta_i^{\text{ext}}(s), \theta_i^*(s)),$$

or $x_i^{\text{ext}}(s) = x_i(\theta_i^*(s))$, $y_i^{\text{ext}}(s) = y_i(\theta_i^*(s))$, $\zeta_i^{\text{ext}}(s) = \zeta_i(\theta_i^*(s))$, etc.

In view of the definitions, we have that

$$\theta_i^*(s) = \int_0^s \alpha_i(\varsigma) d\varsigma, \quad \hat{\theta}(s) = \int_0^s \hat{\alpha}(\varsigma) d\varsigma,$$

and

$$\zeta_i^{\text{ext}}(s) = \int_0^s \beta_i(\varsigma) d\varsigma, \quad \hat{\zeta}^{\text{ext}}(s) = \int_0^s \hat{\beta}(\varsigma) d\varsigma.$$

Therefore, by taking into account Exercise 5.7, and by selecting a subsequence, we have $\alpha_i(s) \xrightarrow{w} \hat{\alpha}(s)$, $|\beta_i(s)| \xrightarrow{w} |\hat{\beta}(s)|$, and $\beta_i(s) \xrightarrow{w} \hat{\beta}(s)$ weakly in L_2. Note that it is not restrictive to consider that $\alpha_i(s) \to 0$ for a.a. s: $\hat{\alpha}(s) = 0$ as α_i are nonnegative. From Exercise 5.7, we also have $x_i^{\text{ext}} \rightrightarrows \hat{x}^{\text{ext}}$.

From (5.19) and (5.22), it follows that the functions $\psi_i^{\text{ext}}(s)$ are equicontinuous and uniformly bounded. The function $\eta_i^{\text{ext}}(s)$ is decreasing, $\eta_i^{\text{ext}}(1) = 0$, while the value $\eta_i^{\text{ext}}(0)$ is exactly the total variation of η_i^{ext}. Thus, by means of the Arzela–Ascoli theorem and of the Helly selection theorem, and in view of (5.22), we obtain λ, ψ^{ext}, ω and η^{ext}, such that, after extracting a subsequence, $\lambda_i \to \lambda$, $\psi_i^{\text{ext}} \rightrightarrows \psi^{\text{ext}}$, $\omega_i \to w$, $\eta_i^{\text{ext}}(s) \to \eta^{\text{ext}}(s)$, a.a. $s \in [0, 1]$, where $\eta^{\text{ext}}(s)$ is some nonnegative and left continuous on $(0, 1)$, and decreasing function such that $\eta^{\text{ext}}(1) = 0$.

Let us show that the functions σ_i and ξ_i converge uniformly to zero. Regarding σ_i, this fact is obvious since $\zeta_i - \bar{\zeta}_i \rightrightarrows 0$, due to $\|V_i\|_{\mathrm{L}_1} \to 0$ and also to Condition (5.22). In what concerns the function ξ_i, it is enough to show that $\xi_i(0) \to 0$.

Clearly,

$$\xi_i(0) = \int_0^1 3\lambda_i \frac{\chi_i^{+2}(\theta_i^*(s))}{z_i^4} ds = -\int_0^1 \frac{3\chi_i^+(\theta_i^*(s))}{2z_i} d\eta_i^{\text{ext}}(s).$$

Therefore, it is sufficient to show that $\dfrac{\chi_i^+(\theta_i^*(s))}{z_i} \rightrightarrows 0$. If we assume the opposite, then there exist a number $\varepsilon > 0$ and a sequence $s_i \in [0, 1]$ such that $\chi_i^+(\theta_i^*(s_i)) \geq 2\varepsilon z_i, \forall i$. Note that the functions $\chi_i^+(\theta_i^*(s))$ are Lipschitz continuous uniformly in i. Then, there exists a constant $\delta > 0$ such that

$$\chi_i^+(\theta_i^*(s)) \geq \varepsilon z_i, \quad \forall s \in O_i = [s_i - \delta z_i, s_i + \delta z_i].$$

From here, it follows that

$$\int_0^1 \chi_i^{+2}(\theta_i^*(s)) z_i^{-3} ds \geq \int_{[0,1]\cap O_i} \chi_i^{+2}(\theta_i^*(s)) z_i^{-3} ds \geq \int_{[0,1]\cap O_i} \varepsilon^2 z_i^{-1} ds \geq \delta\varepsilon^2 > 0,$$

which, obviously, contradicts (5.16). Thus, $\dfrac{\chi_i^+(\theta_i^*(s))}{z_i} \rightrightarrows 0$ and, then, $\xi_i(0) \to 0$.

By passing to the limit in (5.19), we obtain, as $i \to \infty$, Eq. (5.11). (As the passage to limit is evident, the details have been omitted. See Exercise 5.7.)

The passage to limit in the transversality condition (5.20) is also standard. For this, use is made of the fact that $\xi_i(0) \to 0$, the fact that, in view of (5.16), $z_{0,i}^{-1} \operatorname{dist}(p_i, S) \to 0$, the results on subdifferentiation of the distance function gathered in [31] (Sect. 1.3.3, see Theorems 1.97 and 1.105 therein), the upper semicontinuity of the cone $N_S(p)$, and also (5.22). Then, as $i \to \infty$, (5.12) follows.

Let us prove (5.14). By virtue of (5.17), one has

$$\int_0^1 \frac{|v_i(\theta_i^*(s)) - \bar{v}_i(\theta_i^*(s))|}{m_i(\theta_i^*(s))} ds = a_i^{-1} \int_{t_0}^{t_1} |v_i(t) - \bar{v}_i(t)| dt \to 0.$$

By using this, according to the chosen number κ_i, we conclude that

$$\ell(A) = 1, \text{ where } A = \left\{ s \in [0, 1] : \liminf_{i\to\infty} \kappa_i^{-1} |v_i(\theta_i^*(s))| = 0 \right\}.$$

Therefore, for almost all s there exists a subsequence such that

$$\kappa_i^{-1} |v_i(\theta_i^*(s))| \to 0.$$

Take $s \in [0, 1]$ satisfying this convergence property and an arbitrary vector $v \in K \cap B_{\mathbb{R}^k}$. Then, by selecting an appropriate subsequence, it follows that $\kappa_i \alpha_i(s) \to \infty$. By substituting in the left-hand side of (5.21) the vector $\kappa_i v$, and by dividing both parts of (5.21) by $\kappa_i \alpha_i(s)$, we obtain, as $i \to \infty$,

$$\langle \bar{Q}^{\text{ext}}(s), v \rangle - \omega |v| \leq 0. \tag{5.23}$$

Here, use was also made of the above-proven fact that $\sigma_i(s) \rightrightarrows 0$.

By substituting on the left-hand side of (5.21), the values $u = u_i^{\text{ext}}(s)$ and $v = 0$, and after integrating and by passing to the limit as $i \to \infty$, we obtain

$$\int_0^1 \left(\langle \bar{Q}^{\text{ext}}(s), \hat{\beta}(s) \rangle - \omega |\hat{\beta}(s)| \right) ds \geq 0.$$

Therefore, from Condition (5.23), which holds for every $v \in K$, it is simple to extract

$$\langle \bar{Q}^{\text{ext}}(s), \hat{\beta}(s) \rangle - \omega |\hat{\beta}(s)| = 0 \quad \text{a.a. } s \in [0, 1].$$

Thus, Condition (5.14) is proved.

To prove (5.13), we set $v = 0$ in (5.21). Fix any measurable function $u(s) : [0, 1] \to U$. Take $u = u(s)$ on the left-hand side of (5.21), and integrate both parts of (5.21). After taking the limit as $i \to \infty$, which is standard, and using Condition (5.14) already established above, we conclude that

$$\int_0^1 \hat{\alpha}(s) \bar{H}^{\text{ext}}(u(s), s) ds \leq \int_0^1 \hat{\alpha}(s) \bar{H}^{\text{ext}}(s) ds.$$

However, this is precisely Condition (5.13), though derived in the equivalent integral form. Thus, (5.13) is proved.

Note that, if $\lambda_i \to 0$, $\psi_i^{\text{ext}} \rightrightarrows 0$, $\omega_i \to 0$, and $\eta_i^{\text{ext}}(0) \to 0$, then, inevitably, $\rho_i \to 0$. Indeed, this is due to (5.20), the definition of ρ_i, and also Theorem 1.105 from [31], by means of which it holds that $|h| = 1$ when $h \in \partial \operatorname{dist}(p, S)$ and $p \notin S$. Therefore, nontriviality condition (5.10) clearly follows from (5.20).

Properties (a)–(c) and (a′)–(c′) are satisfied by construction.

This completes the proof under the temporary assumption made at the outset that set U is bounded. In the case of unbounded U, consider the intersection $U_\gamma := U \cap \gamma B_{\mathbb{R}^m}$, where $\gamma > 0$ is sufficiently large. By applying the above-obtained results to the problem with the bounded set U_γ and, then, by passing to the limit as $\gamma \to \infty$ in the maximum principle for bounded problems, we prove the theorem in its full generality. □

5.4 Nondegeneracy Conditions

This section deals with the phenomena of degeneracy of the maximum principle. This important phenomenon in its various aspects has been studied in [6, 14, 32–42] and in some other work. It follows that the maximum principle in Theorem 5.1 degenerates in many interesting cases. Let us illustrate what the term "degeneration" means in the current context. Let $d = 1$, left endpoint x_0 be fixed, and $l(x_0, t_0) = 0$. Then, the trivial, albeit non-zero, set of Lagrange multipliers

$$\lambda = 0,\ \omega = 0,\ \psi = 0,\ \psi_\tau = 0,\ \eta = 0,\ \eta_{t_0} = \begin{cases} 1, & s = 0, \\ 0, & s > 0, \end{cases},\ \eta_\tau = 0,\ \ \forall \tau > t_0 \tag{5.24}$$

satisfies Theorem 5.1 if $t_0 \in \mathrm{Ds}(\hat{\mu})$.

If $t_0 \notin \mathrm{Ds}(\hat{\mu})$, the Lagrange multipliers

$$\lambda = 0,\ \omega = 0,\ \psi = 0,\ \psi_\tau = 0,\ \eta = \begin{cases} 1, & t = t_0, \\ 0, & t > t_0, \end{cases},\ \eta_\tau = 0 \tag{5.25}$$

trivially satisfy the maximum principle as well. Thus, the assertion of the maximum principle becomes meaningless.

Therefore, an obvious question arises: How to obtain an appropriate nontriviality condition, stronger than (5.4), in order to avoid this degeneracy phenomenon? Let us answer this question.

Definition 5.3 The state constraints are said to be regular provided that, for any (x, t) such that $l(x, t) \le 0$, there exists a vector $z = z(x, t)$ satisfying

$$\left\langle \frac{\partial l^j}{\partial x}(x, t), z \right\rangle > 0, \quad \forall\, j \text{ s.t. } l^j(x, t) = 0.$$

Definition 5.4 The state constraints are said to be compatible with the endpoint constraints at point $\hat{p}$ provided that there exists $\varepsilon > 0$ such that

$$S \cap (\hat{p} + \varepsilon B_{\mathbb{R}^{2n+1}}) \subseteq \{p = (x_0, x_1, y_1) : \ l(x_0, t_0) \le 0,\ l(x_1, t_1) \le 0\}.$$

Definition 5.5 The controllability conditions w.r.t. the state constraints are said to be fulfilled provided that, for $s = 0, 1$,

$$\exists\, \alpha_s \in [0, 1],\ \ f_s \in \operatorname{conv} f(\hat{x}_s, U, t_s),\ \ g_s \in G(\hat{x}_s, t_s)(K \cap S_{\mathbb{R}^k}) \text{ s.t.,}$$
$$(-1)^s \left[\left\langle \frac{\partial l^j}{\partial x}(\hat{x}_s, t_s), \alpha_s f_s + (1 - \alpha_s) g_s \right\rangle + \alpha_s \frac{\partial l^j}{\partial t}(\hat{x}_s, t_s) \right] < 0,$$

$\forall\, j$ such that $l^j(\hat{x}_s, t_s) = 0$. Here, conv denotes the convex hull of a set and $S_{\mathbb{R}^k}$ stands for the simplex in $\mathbb{R}^k$.

Theorem 5.2 *Let Hypothesis (H2) be in force and the collection* $(\hat{x}, \hat{y}, \hat{u}, \hat{\vartheta})$, *where* $\hat{\vartheta} = (\hat{\mu}; \{\hat{v}_\tau\})$, *be an optimal process in Problem* (5.2). *Assume that the state constraints are regular and compatible with the endpoint constraints at point* $\hat{p}$, *while the controllability conditions w.r.t. the state constraints are fulfilled.*

Then, for any Lagrange multipliers $(\lambda, \omega, \psi, \psi_\tau, \eta, \eta_\tau)$ *associated with the process* $(\hat{x}, \hat{y}, \hat{u}, \hat{\vartheta})$, *the following additional conditions to the maximum principle stated in Theorem 5.1 are satisfied:*

$$\lambda + |\omega| + \ell\left\{t \in T : \left(\psi(t) - \frac{\partial l}{\partial x}(t)\eta(t)\right) \neq 0\right\} + \sum_{\tau \in \mathrm{Ds}(\hat{\mu})} \hat{v}_\tau \cdot \ell\left\{s \in [0,1] : \left(\psi_\tau(s) - \frac{\partial l_\tau}{\partial x}(s)\eta_\tau(s)\right) \neq 0\right\} > 0, \tag{5.26}$$

$$h(t) = h(t_0) + \int_{t_0}^{t} \frac{\partial \bar{H}}{\partial t}(\varsigma)d\varsigma + \int_{[t_0,t]} \frac{\partial}{\partial t}\langle \bar{Q}(\varsigma), d\hat{\mu}_c\rangle + \sum_{\tau \in \mathrm{Ds}(\hat{\mu}):\, \tau \leq t} \left(h_\tau(1) - h(\tau^-)\right), \quad \forall t \in (t_0, t_1]$$

$$\begin{cases} \dot{h}_\tau(s) = \dfrac{\partial}{\partial t}\langle \bar{Q}_\tau(s), \hat{v}_\tau(s)\rangle, \\ h_\tau(0) = h(\tau^-), \end{cases} \quad s \in [0,1], \ \forall \tau \in \mathrm{Ds}(\hat{\mu})$$

where $h(t) = \max_{u \in U}\left\{\bar{H}(u,t)\right\}$, $\forall t \in T$.

The new nontriviality condition (5.26) forbids trivial Lagrange multipliers such as those in (5.24), (5.25).

The discontinuous variable change and the results from [6] may be used to prove this result. Another possible approach could be as in [1], where nondegeneracy conditions for state-constrained optimal impulsive control problems have also been derived. However, the ones in this chapter ensure the nondegeneracy of the necessary optimality condition in the Gamkrelidze-like form. For the phenomenon of degeneracy in conventional optimal control problems, and also for the background to the question, we refer the reader to monograph [14] as well as to the bibliography cited therein.

Let us demonstrate that the assumption of controllability from Definition 5.4 is essential and cannot be omitted. Consider the following example (see Example 4.2 in [1]).

Example 5.1 Consider the problem

$$\begin{aligned} &\text{Minimize} \int_{[0,1]} d\mu \\ &\text{subject to: } dx = 2t d\mu, \\ &\qquad x(t) \ge t^2, \\ &\qquad x_0 = 0, \ x_1 = 1, \\ &\qquad \mu \ge 0. \end{aligned}$$

In [1], it is shown that the minimizer is the Lebesgue measure: $\mu^* = \ell$. We have

$$\bar{Q}(t) = 2t(\psi(t) + \eta(t)) - \lambda.$$

Since $\mathrm{supp}(\mu^*) = [0, 1]$, it follows from (5.9) that $\bar{Q}(t) = 0 \ \forall\, t \in [0, 1]$.

Clearly, if $\lambda = 0$, then $\psi(t) + \eta(t) = 0 \ \forall\, t \in (0, 1]$, which contradicts (5.26). Thus, $\lambda \neq 0$, and the function $\psi(t) + \eta(t)$ is unbounded in the proximity of point $t = 0$. The latter, however, is not possible due to the fact that both ψ and η are bounded. This contradiction is due to the violation of the controllability condition at $t = 0$.

5.5 Exercises

Exercise 5.1 Let (x, u, ϑ), $\vartheta = (\mu; \{v_\tau\})$ be a control process, i.e.,

$$dx = f(x, u, t)dt + G(x, t)d\vartheta, \quad t \in T, \quad x(t_0) = x_0. \tag{5.27}$$

Demonstrate that there exists a sequence of nonimpulsive (conventional) controls $v_i \in \mathrm{L}_\infty(T; \mathbb{R}^k)$, such that $v_i(t) \in K \subseteq \mathbb{R}^k_+$, a.a. t, $\mu_i \overset{w*}{\to} \mu$, and solution x_i to the differential equation (5.27) in which ϑ is replaced by μ_i exists on T. Moreover, $x_i \circ \theta_i \rightrightarrows x^{\mathrm{ext}}$. Here, $\frac{d\mu_i}{dt} = v_i(t), \theta_i$ is the inverse function to the time variable change defined by measure μ_i, that is, to the function:

$$\pi_i(t) = \frac{t - t_0 + |\mu_i|([t_0, t])}{t_1 - t_0 + \|\mu_i\|}, \tag{5.28}$$

and $x^{\mathrm{ext}} : [0, 1] \to \mathbb{R}^n$ is the extended trajectory x w.r.t. ϑ. Use only the assumption imposed in H1, without the assistance of Estimate (4.7) in Chap. 4.

Exercise 5.2 Prove an analogue of Lemma 4.1 in Chap. 4. Show that, under H1, Condition (4.7) can be replaced by the condition that $x_i(\cdot)$'s are uniformly bounded w.r.t. C-metric.

Exercise 5.3 Show that the set $\mathcal{M}$ constructed in the proof of Theorem 5.1 is indeed closed. Take into account Exercise 5.2.

Exercise 5.4 Ensure that function $\Delta(\alpha, \beta)$ introduced in the proof of Theorem 5.1 is lower semicontinuous.

Exercise 5.5* Let $\vartheta = (\mu; \{v_\tau\})$ be impulsive control and $d\zeta = d\vartheta$. Let the sequence of conventional controls $\bar{v}_i \in L_\infty(T; \mathbb{R}^k)$ converges to ϑ, albeit, in the extended sense, that is, $F(\cdot; \bar{\mu}_i) \circ \bar{\theta}_i \rightrightarrows \zeta^{\text{ext}}$, where $d\bar{\mu}_i = \bar{v}_i(t)dt$, $\bar{\theta}_i$ is the inverse to the time variable change (5.28) defined by $\bar{\mu}_i$, and the extension ext is considered w.r.t. ϑ.

Let

$$\int_{t_0}^{t_1} |\zeta_i(t) - \bar{\zeta}_i(t)|^2 \cdot m_i(t)dt \to 0,$$

where

$$\dot{\zeta}_i = v_i(t), \quad m_i(t) := 1 + \frac{|v_i(t)| + |\bar{v}_i(t)|}{2},$$

while $\{v_i\}$ is some other sequence of conventional controls in $L_\infty(T; \mathbb{R}^k)$. Then, $\theta_i \rightrightarrows \theta$, and $F(\cdot; \mu_i) \circ \theta_i \rightrightarrows \zeta^{\text{ext}}$, where $d\mu_i = v_i(t)dt$ and θ_i, θ are the inverse to the time variable change (5.28) defined by μ_i and μ, respectively.

Exercise 5.6 Let ϑ and v_i be as defined in Exercise 5.5. Let $x_{0,i} \to x_0, u_i \to u$ in L_1 and x, x_i be trajectories corresponding to (x_0, u, ϑ) and $(x_{0,i}, u_i, v_i)$, respectively, by means of (5.27). Then, $x_i \circ \theta_i \rightrightarrows x^{\text{ext}}$, where $x^{\text{ext}} : [0, 1] \to \mathbb{R}^n$ is the extended trajectory x w.r.t. ϑ.

Exercise 5.7 Ensure that the assertions of Exercises 5.5, 5.6 can be complemented. Prove that $\theta_i^* \rightrightarrows \theta$, $F(\cdot; \mu_i) \circ \theta_i^* \rightrightarrows \zeta^{\text{ext}}$. Here, θ_i^* is the inverse to the time variable change (5.28) defined by μ_i^*, where $d\mu_i^* = m_i(t)dt$. Then, also, $x_i \circ \theta_i^* \rightrightarrows x^{\text{ext}}$.

Exercise 5.8 Construct the degenerate set of multipliers (5.24) and (5.25), but, now, for the right time endpoint $t = t_1$.

References

1. Arutyunov, A., Karamzin, D., Pereira, F.: A nondegenerate maximum principle for the impulse control problem with state constraints. SIAM J. Control Optim. **43**(5), 1812–1843 (2005)
2. Gusev, M.: On optimal control of generalized processes under nonconvex state constraints. Differential Games and Control Problems [in Russian], UNTs, Akad. Nauk SSSR, Sverdlovsk **15**, 64–112 (1975)
3. Miller, B.: Generalized solutions in nonlinear optimization problems with impulse controls. i. the solution existence problem. Avtomatika i Telemekhanika **4**, 62–76 (1995)

4. Miller, B.: Generalized solutions in nonlinear optimization problems with impulse controls. ii. representation of solutions by differential equations with measure. Avtomatika i Telemekhanika **5**, 56–70 (1995)
5. Gamkrelidze, R.: Optimal control processes for bounded phase coordinates. Izv. Akad. Nauk SSSR. Ser. Mat. **24**, 315–356 (1960)
6. Arutyunov, A., Karamzin, D., Pereira, F.: The maximum principle for optimal control problems with state constraints by R.V. Gamkrelidze: revisited. J. Optim. Theory Appl. **149**(3), 474–493 (2011)
7. Neustadt, L.: An abstract variational theory with applications to a broad class of optimization problems. i. general theory. SIAM J. Control **4**(3), 505–527 (1966)
8. Neustadt, L.: An abstract variational theory with applications to a broad class of optimization problems. ii. applications. SIAM J. Control **5**(1), 90–137 (1967)
9. Pontryagin, L., Boltyanskii, V., Gamkrelidze, R., Mishchenko, E.: Mathematical theory of optimal processes. Translation from the Russian ed. by L.W. Neustadt. Interscience Publishers, Wiley, 1st edn (1962)
10. Russak, I.: On general problems with bounded state variables. J. Optim. Theory Appl. **6**(6), 424–452 (1970)
11. Warga, J.: Minimizing variational curves restricted to a preassigned set. Trans. Amer. Math. Soc. **112**, 432–455 (1964)
12. Dubovitskii, A., Milyutin, A.: Extremum problems with constraints. Sov. Math. Dokl. **4**, 452–455 (1963)
13. Dubovitskii, A., Milyutin, A.: Extremum problems in the presence of restrictions. USSR Comput. Math. Math. Phys. **5**(3), 1–80 (1965)
14. Arutyunov, A.V.: Optimality conditions. Abnormal and degenerate problems. In: Mathematics and its Applications. Kluwer Academic Publishers, Dordrecht (2000)
15. Halkin, H.: A satisfactory treatment of equality and operator constraints in the dubovitskii-milyutin optimization formalism. J. Optim. Theory Appl. **6**(2), 138–149 (1970)
16. Ioffe, A., Tikhomirov, V.: Studies in Mathematics and its Applications, vol. 6. Elsevier Science, North-Holland, Amsterdam (1979)
17. Maurer, H.: Differential stability in optimal control problems. Appl. Math. Optim. **5**(1), 283–295 (1979)
18. Milyutin, A.: Maximum Principle for General Optimal Control Problem. Fizmatlit, Moscow (2001). [in Russian]
19. Clarke, F.: Optimization and Nonsmooth Analysis. Wiley-Interscience, New York (1983)
20. Ioffe, A.: Necessary conditions in nonsmooth optimization. Math. Op. Res. **9**(2), 159–189 (1984)
21. Vinter, R., Pappas, G.: A maximum principle for nonsmooth optimal-control problems with state constraints. J. Math. Anal. Appl. **89**(1), 212–232 (1982)
22. Arutyunov, A., Karamzin, D.: Non-degenerate necessary optimality conditions for the optimal control problem with equality-type state constraints. J. Glob. Optim. **64**(4), 623–647 (2016)
23. Davydova, A., Karamzin, D.: On some properties of the shortest curve in a compound domain. Differ. Equ. **51**(12), 1626–1636 (2015)
24. Arutyunov, A.: Properties of the Lagrange multipliers in the pontryagin maximum principle for optimal control problems with state constraints. Differ. Equ. **48**(12), 1586–1595 (2012)
25. Arutyunov, A., Karamzin, D.: On some continuity properties of the measure lagrange multiplier from the maximum principle for state constrained problems. SIAM J. Control Optim. **53**(4), 2514–2540 (2015)
26. Arutyunov, A., Karamzin, D., Pereira, F.: Conditions for the absence of jumps of the solution to the adjoint system of the maximum principle for optimal control problems with state constraints. Proc. Stekl. Inst. Math. **292**, 27–35 (2016)
27. Arutyunov, A., Karamzin, D.: Properties of extremals in optimal control problems with state constraints. Differ. Equ. **52**(11), 1411–1422 (2016)
28. Arutyunov, A., Karamzin, D., Pereira, F.: Investigation of controllability and regularity conditions for state constrained problems. In: Proceedings of the 20-th IFAC Congress, Toulouse, France (2017)

29. Ekeland, I.: On the variational principle. J. Math. Anal. Appl. **47**, 324–353 (1974)
30. Mordukhovich, B.: Variational Analysis and Generalized Differentiation II. Applications. In: Grundlehren der Mathematischen Wissenschaften [Fundamental Principles of Mathematical Sciences], vol. 331. Springer, Berlin (2006)
31. Mordukhovich, B.: Variational Analysis and Generalized Differentiation I. Basic Theory. In: Grundlehren der Mathematischen Wissenschaften [Fundamental Principles of Mathematical Sciences], vol. 330. Springer, Berlin (2006)
32. Arutyunov, A.: On the theory of the maximum principle for state constrained optimal control problems with state constraints. Dokl. Math. SSSR **304**(1) (1989)
33. Arutyunov, A.: Perturbations of extremal problems with constraints and necessary optimality conditions. J. Sov. Math. **54**(6), 1342–1400 (1991)
34. Arutyunov, A., Aseev, S.: Investigation of the degeneracy phenomenon of the maximum principle for optimal control problems with state constraints. SIAM J. Control Optim. **35**(3), 930–952 (1997)
35. Arutyunov, A., Aseev, S., Blagodatskikh, V.: First-order necessary conditions in the problem of optimal control of a differential inclusion with phase constraints. Sb. Math. **79**(1), 117–139 (1994)
36. Arutyunov, A., Tynyanskij, N.: The maximum principle in a problem with phase constraints. Sov. J. Comput. Syst. Sci. **23**(1), 28–35 (1984)
37. Arutyunov, A., Tynyanskiy, N.: Maximum principle in a problem with phase constraints. Sov. J. Comput. Syst. Sci. **23**(1), 28–35 (1985)
38. Dubovitskij, A., Dubovitskij, V.: Necessary conditions for a strong minimum in optimal control problems with degeneration of final and phase constraints. Usp. Mat. Nauk **40**(2(242)), 175–176 (1985)
39. Ferreira, M., Fontes, F., Vinter, R.: Nondegenerate necessary conditions for nonconvex optimal control problems with state constraints. J. Math. Anal. Appl. **233**(1), 116–129 (1999)
40. Ferreira, M., Vinter, R.: When is the maximum principle for state constrained problems nondegenerate? J. Math. Anal. Appl. **187**(2), 438–467 (1994)
41. Palladino, M., Vinter, R.: Regularity of the hamiltonian along optimal trajectories. SIAM J. Control Optim. **53**(4), 1892–1919 (2015)
42. Vinter, R.: Optimal Control. Birkhäuser, Boston (2000)

Chapter 6
Impulsive Control Problems with Mixed Constraints

Abstract In this chapter, two directions for investigation are combined. On the one hand, mixed constraints are added to the formulation of the problem, that is, joint constraints on the state variable and on the control variable. Such constraints are in demand in engineering applications. On the other hand, a new and broader impulsive extension concept is considered, as it is assumed that the matrix-multiplier G may now depend on both the state variable x and the control of the conventional type u. This leads to a new, more general type of impulsive control which can be found in various engineering applications, for example, those in which rapid variations in the mass distribution of a mechanical system need to be taken into account for the small time interval when the impulse takes place. A corresponding model example of such a control system equipped with the mixed constraints is given in Sect. 6.2. Further on in this chapter, the maximum principle is proved which requires some effort and auxiliary techniques contained in Sect. 6.4. The chapter ends with ten exercises.

6.1 Introduction

This chapter, once again, follows the rationale of increasing complexity of the impulsive extension described in the Introduction. More general dynamical control system (0.7) is investigated relative to the extension issue. Its extension involves a more intricate concept of the impulsive control by encompassing extra controls u_τ, which are conventional bounded controls. These act on the jumps of the impulsive control system.

This type of extended impulsive control can be encountered in various engineering applications in which, for example, it is necessary to take into account rapid variations in mass distribution of a mechanical system during the short time when the impulse is applied. There are, of course, many other applications. In the next section, we provide a detailed example showing how these controls can be useful.

© Springer Nature Switzerland AG 2019
A. Arutyunov et al., *Optimal Impulsive Control*, Lecture Notes in Control and Information Sciences 477, https://doi.org/10.1007/978-3-030-02260-0_6

Here, we also derive the Pontryagin maximum principle for impulsive control problems with a new type of constraint, the so-called geometric mixed constraints. The mixed constraints of geometric type are given by the following relation

$$r(x, u, t) \in C,$$

where r is some function smooth w.r.t. x, u, and C is a closed set. The fact that both variables, state x and control u, are related to each other, as the value of one variable conditions the value of the other through the joint constraint, gives rise to the term "mixed." The key issue in the investigation of mixed constraints is the so-called regularity condition or constraint qualification. If C is merely closed, such a condition is expressed in terms of the limiting normal cone to the set, [16], henceforth referred to as Robinson constraint qualification (RCQ) condition, [24]. The classic maximum principle (with or without impulsive controls) is true only under RCQ imposed globally on the feasible set. If RCQ is not satisfied, then the maximum condition may not hold, and, in consequence, extensions involving more general forms of the maximum principle are required under some weakened regularity assumptions (see, e.g., in [3, 4]).

Although there is extensive literature on the topic of conventional optimal control problems with mixed constraints (see, e.g., [1, 2, 5–9, 12–14, 18–22], and the bibliography cited therein), the impulsive control problems under mixed constraints have previously been investigated to a much weaker extent. Here, some results are derived which help fill this gap. In the absence of mixed constraints, and for the data smooth w.r.t. time, the proof of maximum principle is given in Chap. 4. This proof is rather simple due to a reduction to a conventional optimal control problem. However, in the present chapter, mere measurability of the right-hand side of the dynamics w.r.t. time is assumed. This, and the consideration of mixed constraints, creates new challenges that are overcome by using a technique of proof similar to that in Chap. 5.

The chapter is organized as follows. In Sect. 6.2, an example of a mechanical system is demonstrated, which admits impulses, but which is not addressed in the conventional study of impulsive systems. The example shows the relevance of the principle topic of this chapter: to find a proper extension of a control problem under (0.7). In Sect. 6.3, a detailed statement of the optimal impulsive control problem is presented, including the required assumptions imposed on its data as well as some basic definitions. Regularity of mixed constraints and RCQ are introduced. In Sect. 6.4, some technical constructions and lemmas that play a key role in the derivation of the main result—Theorem 6.1, formulated and proved in Sect. 6.5—are brought into consideration. The chapter concludes with Sect. 6.6, containing useful exercises.

6.2 Example

Consider a controllable disk in $\mathbb{R}^2$ (an imaginary flying object) of unit radius, whose motion between two given points A and B of the plane is carried out by four thrusters located at the points $(1, 0)$, $(0, 1)$, $(-1, 0)$, $(0, -1)$ of unit circle—that is the disk boundary—given in the local (body frame) coordinate system. Let these points be numbered $1, 2, 3, 4$. Assume that the disk rotation is taken into account. For example, we wish to steer the disk keeping the angular velocity equal to some constant c. Let CG denote the center of gravity of the disk which cannot be considered constant due to the fuel consumption. (Fuel contributes to the disk mass and is such that it changes the mass distribution of the disk.) Therefore, the CG position is considered as some function of the disk mass m.

Let us say that the thrust angle is the acute angle of the line of the propulsive force produced by the thruster with the corresponding axis. For instance, for thruster number 1, this angle is between the line of the propulsive force and the axis $0x$, for thruster number 2, the one between the line of the propulsive force and the axis $0y$, etc. We control the thrust angle $\theta_i = \theta_i(t)$, the relative power output $p_i = p_i(t)$ of each thruster, and the fuel consumption rate $v = v(t)$, which is a nonnegative scalar. The controls $p_i \geq 0$, $i = 1, 2, 3, 4$, are such that

$$p_1 + p_2 + p_3 + p_4 = 1.$$

The control angle θ is positive if the thruster is inclined clockwise.

Denote by $\theta_i(m)$ the deviation angle of thruster number i when this thruster is directed to the CG; by $r_i(m)$ the distance between thruster i and the CG; by $i(m)$ the moment of inertia of the disk w.r.t. the CG; by α the orientation angle specified by the position of thruster number 1; by ω the angular velocity w.r.t. the CG. The positive rotation is counterclockwise. The thrusters are schematically shown in Fig. 6.1.

The equations of motion are

$$\begin{aligned}
\dot{x}_1 &= w_1, \\
\dot{x}_2 &= w_2, \\
\dot{w}_1 &= (k_1(m, \theta, p)\cos\alpha - k_2(m, \theta, p)\sin\alpha) \cdot v/m, \\
\dot{w}_2 &= (k_1(m, \theta, p)\sin\alpha + k_2(m, \theta, p)\cos\alpha) \cdot v/m - g, \\
\dot{m} &= -v,
\end{aligned}$$

where $x = (x_1, x_2)$, $w = (w_1, w_2)$, m, g are, respectively, the position, velocity, mass, and the gravity acceleration. The coefficients k_1 and k_2 are given by

$$\begin{aligned}
k_1(m, \theta, p) = {} & p_1 \cos\theta_1(m)\cos(\theta_1 - \theta_1(m)) - p_2 \sin\theta_2(m)\cos(\theta_2 - \theta_2(m)) - \\
& p_3 \cos\theta_3(m)\cos(\theta_3 - \theta_3(m)) + p_4 \sin\theta_4(m)\cos(\theta_4 - \theta_4(m)), \\
k_2(m, \theta, p) = {} & -p_1 \sin\theta_1(m)\cos(\theta_1 - \theta_1(m)) - p_2 \cos\theta_2(m)\cos(\theta_2 - \theta_2(m)) \\
& + p_3 \sin\theta_3(m)\cos(\theta_3 - \theta_3(m)) + p_4 \cos\theta_4(m)\cos(\theta_4 - \theta_4(m)).
\end{aligned}$$

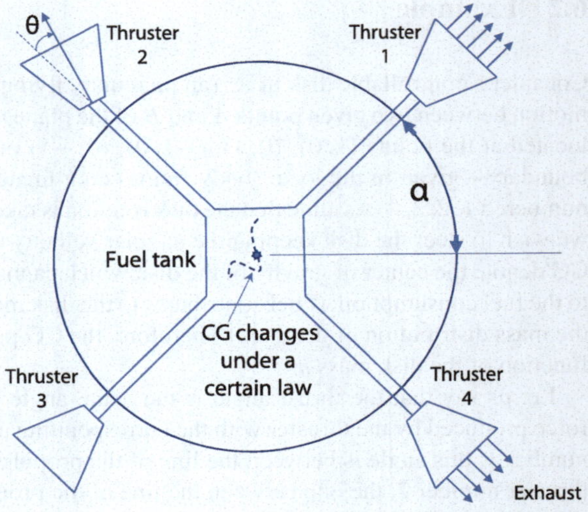

Fig. 6.1 The imaginary reaction control system. The case when CG is in the center of the disk is shown. (Then, $\theta_i(m) = 0$)

It is assumed above that the absolute value of the propulsion force is proportional to the fuel consumption rate and also that the fuel consumption rate can take *unbounded* values. These assumptions are a certain simplification of the reality, but they are natural in thruster control modeling in space.

The sum of torques τ applied by the forces is

$$\tau = \sum_{i=1}^{4} p_i r_i(m) \sin(\theta_i - \theta_i(m)) \cdot v.$$

For the differential of the angular momentum, we have

$$dL = i(m + dm)d\omega + i(m + dm)\omega - \sum_{i=1}^{4} p_i r_i^2(m)\omega dm - i(m)\omega.$$

By the conservation of angular momentum, it follows that $dL = \tau dt$. Then, by using $\dot{m} = -v$, we obtain the following equation for the orientation angle α

$$\dot{\alpha} = \omega,$$
$$\dot{\omega} = \frac{\omega(i'(m) - \sum_i p_i r_i^2(m)) + \sum_i p_i r_i(m) \sin(\theta_i - \theta_i(m))}{i(m)} \cdot v,$$

where $i'(m) = di/dm$, the derivative w.r.t. mass.

Now, the problem of optimal fuel consumption for the controllable disk, in the framework of Sect. 6.3, can be formulated as follows:

$$
\begin{aligned}
\text{Minimize } & m(0) \\
\text{subject to: } & \dot{x}_1 = w_1, \\
& \dot{x}_2 = w_2, \\
& x(0) = A, \ x(1) = B, \\
& dw_1 = \frac{k_1(m,\theta,p)\cos\alpha - k_2(m,\theta,p)\sin\alpha}{m} \cdot d\vartheta, \\
& dw_2 = -g \cdot dt + \frac{k_1(m,\theta,p)\sin\alpha + k_2(m,\theta,p)\cos\alpha}{m} \cdot d\vartheta, \\
& w(0) = w_A, \ w(1) = w_B, \\
& \dot{\alpha} = \omega, \\
& d\omega = \frac{\omega(i'(m) - \sum_i p_i r_i^2(m)) + \sum_i p_i r_i(m)\sin(\theta_i - \theta_i(m))}{i(m)} \cdot d\vartheta, \\
& \alpha(0) = \alpha_A, \ \alpha(1) = \alpha_B, \\
& \omega(0) = \omega_A, \ \omega(1) = \omega_B, \\
& dm = -d\vartheta, \\
& m(1) = M, \\
& \textstyle\sum_i p_i = 1, \ p_i \geq 0, \ |\theta_i| \leq \pi/2, \\
& \vartheta = (\{(\theta_\tau, p_\tau)\}|_{\tau\in[0,1]}, \mu), \ \mu \geq 0,
\end{aligned}
$$

where ϑ is the impulsive control in its new and extended form (see Definition 6.1). If we withdraw the impulsive controls here, by changing $d\vartheta$ into vdt, and thus considering usual measurable controls, then this problem will not have a solution in the class of absolutely continuous trajectories, due to the unbounded scalar control v. Therefore, impulsive controls help to provide a rigorous mathematical description of the actual motion of the disk.

Consider translation of the disk w.r.t. a constant angular velocity, $\omega = \text{const}$. Then, $r(m, \theta, p) = 0$, where

$$
r(m, \theta, p) = \omega\Big(i'(m) - \sum_{i=1}^{4} p_i r_i^2(m)\Big) + \sum_{i=1}^{4} p_i r_i(m)\sin(\theta_i - \theta_i(m)).
$$

This is the situation in which the mixed constraints introduced in this chapter begin to have a clear physical meaning. If $\omega = 0$, or the mass distribution is such that $i'(m) = \sum_i p_i r_i^2(m)$ (which corresponds to the case of entire mass be concentrated at the disk boundary), then, for any optimal control mode, all the force lines intersect the CG, because the fuel is used merely for translation. However, as soon as the angular velocity is not zero and some part of the mass is, for instance, concentrated at the CG, then, according to the mixed constraints, we have to use the thrusters also in the orientation mode, in order to compensate the angular momentum of the exhaust, and thus, to keep the angular velocity constant. Then, the mixed constraints also become meaningful w.r.t. the impulses and the discontinuities of the control system, where they restrict conventional controls during the impulsive control action.

It has been demonstrated above how nontrivial conventional controls may arise as the impulse develops. The next sections are devoted to the study of how to take into account these additional conventional controls attached to the jumps in the optimality conditions under mixed constraints.

6.3 Problem Formulation and Basic Definitions

Consider the following impulsive control problem:

$$\begin{aligned} &\text{Minimize} \quad \varphi(p) \\ &\text{subject to:} \quad dx = f(x,u,t)dt + G(x,u,t)d\vartheta, \\ &\qquad p = (x_0, x_1) \in S, \ t \in T, \\ &\qquad r(x,u,t) \in C, \\ &\qquad \text{range}(\vartheta) \subset K. \end{aligned} \tag{6.1}$$

Here, $T = [t_0, t_1]$ is a fixed time interval, $p = (x_0, x_1)$, where $x_0 = x(t_0), x_1 = x(t_1)$, is the endpoint vector, S is a closed set in $\mathbb{R}^{2n}$, C is a closed set in $\mathbb{R}^d$, K is a closed convex cone embedded in $\mathbb{R}^k_+$, and $\vartheta = (\mu; \{u_\tau\}, \{v_\tau\})$ is a new object which is called impulsive control.

The impulsive control comprises several components. First component μ, as usual, is a vector Borel measure with range in K. The remaining two components, that is $\{u_\tau\}$ and $\{v_\tau\}$, are infinite families of measurable bounded functions defined on interval $[0, 1]$. They depend on real parameter $\tau \in T$. The exact properties of these families as well as the strict definition of the impulsive control and the concept of the solution $x(\cdot)$ are presented later in this section. Function $u(\cdot)$ is called conventional control, and it is assumed to be measurable and essentially bounded w.r.t. both Lebesgue ℓ and Lebesgue–Stieltjes $|\mu|$ measures. Then, the pair (u, ϑ) is called control in problem (6.1). The constraints $r(x, u, t) \in C$ are called mixed. They should be understood in a broader meaning than that of the conventional inclusion. This is due to the presence of impulsive controls. We describe properties of these constraints below.

The functions in (6.1), i.e., $\varphi : \mathbb{R}^{2n} \to \mathbb{R}^1$, $f : \mathbb{R}^n \times \mathbb{R}^m \times \mathbb{R}^1 \to \mathbb{R}^n$, $G : \mathbb{R}^n \times \mathbb{R}^m \times \mathbb{R}^1 \to \mathbb{R}^n \times \mathbb{R}^k$, and $r : \mathbb{R}^n \times \mathbb{R}^m \times \mathbb{R}^1 \to \mathbb{R}^d$ satisfy the following basic hypothesis. The function f and its partial derivatives w.r.t. x, u are Lebesgue measurable w.r.t. t for all x, u and, on any bounded set, continuous w.r.t. x, u uniformly w.r.t. x, u, t and bounded. Matrix-valued function G and vector-valued function r are continuous in all arguments and continuously differentiable w.r.t. x, u. Function φ is continuously differentiable.

Consider a Borel vector measure μ such that $\text{range}(\mu) \subset K$. Take a number $\tau \in T$. Denote by $W_\tau(\mu) \subset L_\infty([0,1]; \mathbb{R}^k)$ the set of functions $v : [0,1] \to K$ satisfying the following two conditions:

(i) $\sum_{j=1}^{k} v^j(s) = \sum_{j=1}^{k} \mu^j(\tau)$, a.a. $s \in [0,1]$;

(ii) $\int_0^1 v^j(s)ds = \mu^j(\tau)$, $j = 1, \dots, k$.

Here, $\mu^j(\tau) = \mu^j(\{\tau\})$ is the value of μ^j at the single point set $\{\tau\}$, and the symbol a.a. s means "almost all s." Note that $W_\tau(\mu) = \{0\}$ whenever $\mu^j(\tau) = 0$, $\forall\, j$.

Definition 6.1 The triple $\vartheta = (\mu; \{u_\tau\}, \{v_\tau\})$ is said to be impulsive control, if $v_\tau \in W_\tau(\mu)$ $\forall\, \tau \in T$, and $\{u_\tau\}$ is a family of measurable essentially bounded uniformly w.r.t. τ vector-valued functions defined on interval $[0, 1]$ with values in $\mathbb{R}^m$. The families $\{u_\tau\}$, $\{v_\tau\}$ are called attached to the vector-valued measure μ. The measure $|\mu|$ is called variation of the impulsive control ϑ and is denoted by $|\vartheta|$.

Take control functions $w \in \mathrm{L}_\infty([0, 1]; \mathbb{R}^m)$, $v \in \mathrm{L}_\infty([0, 1]; \mathbb{R}^k)$, a point $\tau \in T$, and a vector $a \in \mathbb{R}^n$. Denote by $z(\cdot) = z(\cdot; \tau, a, w, v)$ the solution to the following dynamical system

$$\begin{cases} \dot{z}(s) = G(z(s), w(s), \tau)v(s), \quad s \in [0, 1], \\ z(0) = a. \end{cases}$$

A function of bounded variation $x(\cdot)$ on interval T is called solution to the differential equation in (6.1), corresponding to control (u, ϑ), where $\vartheta = (\mu; \{u_\tau\}, \{v_\tau\})$, and initial value x_0, if $x(t_0) = x_0$ and, for all $t \in (t_0, t_1]$,

$$x(t) = x_0 + \int_{t_0}^{t} f(x, u, \tau)d\tau + \int_{[t_0,t]} G(x, u, \tau)d\mu_c + \sum_{\tau \in \mathrm{Ds}(\mu):\, \tau \le t} \Big(x_\tau(1) - x(\tau^-)\Big). \tag{6.2}$$

Here, $x_\tau(\cdot) = z(\cdot; \tau, x(\tau^-), u_\tau, v_\tau)$, μ_c, as before, stands for the continuous component of μ. Note that the sum in (6.2) is well defined because the set of atoms of a Borel measure is countable.

To complete the statement of the problem, it remains to give a meaning to the mixed constraints in (6.1) satisfied by the pair trajectory and control. Take any triple (x, u, ϑ), where trajectory $x(\cdot)$ corresponding to (u, ϑ) is defined as above. Inclusion $r(x, u, t) \in C$ is understood in the following generalized sense:

$$r(x, u, t) \in C \quad \Leftrightarrow \quad \begin{cases} r(x(t), u(t), t) \in C \;\; \text{a.a.}\, t \in T, \\ r(x_\tau(s), u_\tau(s), \tau) \in C, \;\; \text{a.a.}\, s \in [0, 1], \;\; \forall\, \tau \in \mathrm{Ds}(|\vartheta|). \end{cases}$$

The triple (x, u, ϑ) is called control process if (6.2) is satisfied. A control process is said to be admissible if it satisfies all the constraints of problem (6.1). An admissible process $(\hat{x}, \hat{u}, \hat{\vartheta})$ is said to be optimal if, for any admissible process (x, u, ϑ), the inequality $\varphi(\hat{p}) \le \varphi(p)$ is satisfied, where $\hat{p} = (\hat{x}(t_0), \hat{x}(t_1))$.

Our main objective in this chapter is to derive necessary optimality conditions for problem (6.1) in the form of the Pontryagin maximum principle, [23]. In order to ensure nondegeneracy of the maximum principle (the optimality conditions should remain informative; i.e., they should not be satisfied by a trivial set of Lagrange multipliers), it is necessary to use some concept of regular trajectories w.r.t. the mixed constraints. Let us describe this type of regularity.

Consider the set-valued map

$$U(x,t) := \{u \in \mathbb{R}^m : r(x,u,t) \in C\}.$$

Definition 6.2 A point $u \in U(x,t)$ is said to be regular w.r.t. the mixed constraints provided that

$$N_C(r(x,u,t)) \cap \ker \frac{\partial r^*}{\partial u}(x,u,t) = \{0\}. \tag{6.3}$$

Here, the set $N_C(y)$, as usual, designates the limiting normal cone [15], and A^* stands for the conjugate matrix to A. The regularity of the point u means that the so-called Robinson constraint qualification (RCQ) holds at u for the constraint system $r(x,u,t) \in C$, [24]. See also in [16, 17].

The simplest example in which any feasible point $u \in U(x,t)$ is regular is given by the so-called geometric constraint

$$u \in C,$$

where the set C is an arbitrary closed set. Indeed, here, we have: $\frac{\partial r}{\partial u}(x,u,t) = E$, and, then, (6.3) is satisfied for all u, because $\ker E^* = \{0\}$. The same holds for the more general case, in which the matrix $\frac{\partial r}{\partial u}(x,u,t)$ has full rank for all x, u, t, for example:

$$u + h(x,t) \in C,$$

where $h : \mathbb{R}^n \times \mathbb{R}^1 \to \mathbb{R}^m$ is any function. Some examples, in which (6.3) is not satisfied, are given in Exercises 6.1 and 6.2.

The condition (6.3) can be reformulated in the following way: There exists a number $\varepsilon > 0$ such that

$$\left| y \frac{\partial r}{\partial u}(x,u,t) \right| \geq \varepsilon |y|, \quad \forall\, y \in N_C(r(x,u,t)).$$

The upper bound of all such ε's is also known as *modulus of surjection* of the constraint system $M : r(x,u,t) \in C$. Let us denote the modulus of surjection of an arbitrary given constraint system $V : g(z) \in S$ at point z, by $\operatorname{sur} V(z)$.[1]

---

[1] In the literature, the modulus of surjection is introduced for set-valued maps $F : X \to 2^Y$. If spaces X and Y are finite dimensional, then

$$\operatorname{sur} F(x|y) = \inf\{|x^*| : x^* \in D^*F(x,y)(y^*),\ |y^*| = 1\}.$$

Then, the regularity of the point $u \in U(x, t)$ is equivalent to the relation

$$\text{sur}M(x, u, t) > 0.$$

We denote by $U_{\text{reg}}(x, t)$ the subset of all regular points of $U(x, t)$. The subset of points for which $\text{sur}M(x, u, t) \geq \varepsilon$ is denoted by $U^{\varepsilon}_{\text{reg}}(x, t)$. Note that this set may not be closed. It is clear that

$$U^{\varepsilon}_{\text{reg}}(x, t) \subseteq U_{\text{reg}}(x, t) \subseteq U(x, t) \ \ \forall \varepsilon > 0, \ \text{and}$$
$$U^{\alpha}_{\text{reg}}(x, t) \subseteq U^{\beta}_{\text{reg}}(x, t) \ \ \text{for } \alpha > \beta > 0,$$

and $U^{0}_{\text{reg}}(x, t) = U(x, t)$.

The next definition of regularity invokes the notion of extended trajectory which has been used previously (see Definition 3.2). The pair $(x(\cdot), \{x_\tau(\cdot)\})$ is the extended trajectory provided that $x \in \text{BV}(T; \mathbb{R}^n)$, and $\{x_\tau\}$ stands for a family of arcs from $\text{W}_{1,\infty}([0, 1]; \mathbb{R}^n)$ which depends on real parameter $\tau \in T$. It is also natural to assume that there is only a countable number of nonconstant arcs from this family.

Definition 6.3 The extended trajectory $(x(\cdot), \{x_\tau(\cdot)\})$ is said to be regular w.r.t. the mixed constraints provided there is a number $\varepsilon_0 > 0$ such that

$$U(x(t), t) \subseteq U^{\varepsilon_0}_{\text{reg}}(x(t), t), \ \ \text{a.a. } t \in T,$$

and

$$U(x_\tau(s), \tau) \subseteq U^{\varepsilon_0}_{\text{reg}}(x_\tau(s), \tau), \ \ \text{a.a. } s \in [0, 1] \ \ \forall \tau \in T.$$

Both set-valued maps $N_C(\cdot)$ and $\ker \frac{\partial r^*}{\partial u}(\cdot)$ are upper semicontinuous, and, thereby, the regularity condition also holds in some δ-tube about the extended trajectory.

Consider the case (important for applications) in which the map r is continuous w.r.t. t, while the sets $U(x, t)$ are uniformly bounded. Then, it is a simple matter to see that regularity is satisfied for any admissible extended trajectory if the following condition is true

$$N_C(r(x, u, t)) \cap \ker \frac{\partial r^*}{\partial u}(x, u, t) = \{0\}, \ \ \forall u \in U(x, t), \ \ \forall x, t.$$

Such an a priori sufficient condition is easy to verify for large classes of sets C, e.g., convex or semialgebraic.

One important property which is implied by regularity of mixed constraints can be stated as follows. The set-valued map $U(x, t)$ is continuous w.r.t. x, uniformly w.r.t. t. (See Lemma 6.4 in the next section.)

Here, $D^*F(x, y)$ is the limiting coderivative of F at (x, y), [16]. By definition, $\text{sur}F(x|y) = \infty$ if $y \notin F(x)$. If we set $F(\cdot) := r(x, \cdot, t) - C$, then $\text{sur}M(x, u, t) = \text{sur}F(x, u, t|0)$.

6.4 Basic Constructions and Lemmas

In this section, some preparative material is given for the proof of the main result. As was mentioned in the Introduction chapter, an important question for any extension is the question of metrics in the space of generalized controls. In other words, that is the question of how to define the notion of the closeness of controls in the extended problem. Let us define this notion now.

Consider the set of all controls; i.e., all the pairs $\wp = (u, \vartheta)$. Denote it by $\mathscr{P}$. Take an arbitrary pair $\wp = (u, \vartheta)$, $\vartheta = (\mu; \{u_\tau\}, \{v_\tau\})$. Consider the discontinuous time variable change $\pi : T \to [0, 1]$,

$$\pi(t) = \frac{t - t_0 + |\vartheta|([t_0, t])}{t_1 - t_0 + |\vartheta|(T)}, \ t \in (t_0, t_1], \ \ \pi(t_0) = 0.$$

The inverse function $\theta(s)$: $[0, 1] \to T$ is such that

(a) $\theta(s)$ is monotone increasing function on $[0, 1]$;
(b) $|\theta(s) - \theta(t)| \le |s - t| \ \forall s, t$;
(c) $\theta(s) = \tau, \forall s \in \Gamma_\tau, \forall \tau \in T$, where $\Gamma_\tau = [\pi(\tau^-), \pi(\tau^+)]$.

Function $\pi(t)$ maps $(\ell + |\vartheta|)$-measurable sets into ℓ-measurable sets. Therefore, if a set A is $(\ell + |\vartheta|)$-measurable, then $\theta^{-1}(A)$ is measurable. This implies that $u(\theta(s))$ is measurable if $u(t)$ is $(\ell + |\vartheta|)$-measurable. So, for example, the following integration variable change is possible:

$$\int_{t_0}^{t_1} |u(t)| d\mu = c \int_0^1 |u(\theta(s))| m(\theta(s)) ds,$$

where $c = t_1 - t_0 + |\vartheta|(T)$ and $m(t)$ is the Radon–Nikodym derivative of measure μ w.r.t. $\ell + |\vartheta|$.

Let h be an arbitrary map that depends on variables x, u, t. Let us define on $[0, 1]$ the operator

$$\mathrm{Ext}[h, \vartheta](s) := \begin{cases} h(x_\tau(\xi_\tau(s)), u_\tau(\xi_\tau(s)), \tau), & \text{if } s \in \Gamma_\tau \text{ for some } \tau \in \mathrm{Ds}(\vartheta), \\ h(x(\theta(s)), u(\theta(s)), \theta(s)), & \text{otherwise.} \end{cases}$$

Here, $\xi_\tau(s) = (s - \pi(\tau^-))/\ell(\Gamma_\tau) : \Gamma_\tau \to [0, 1]$.

Operator $\mathrm{Ext}[h, \vartheta]$ is introduced in order to simplify the notation. For example, now, it is clear to see that the control process (x, u, ϑ) satisfies the mixed constraints iff $\mathrm{Ext}[r, \vartheta](s) \in C$ a.a. $s \in [0, 1]$; i.e., the inequality $r(x, u, t) \in C$ is satisfied in the extended sense for the extended mapping. Sometimes, when it is clear with respect to which impulsive control the extension is considered, we will write simply $h^{\mathrm{ext}}(s)$ instead of $\mathrm{Ext}[h, \vartheta](s)$.

Let us equip set $\mathscr{P}$ with some metric now. Consider two elements $\wp_1, \wp_2 \in \mathscr{P}$ such that $\wp_1 = (u_1, \vartheta_1)$, $\wp_2 = (u_2, \vartheta_2)$, where $\vartheta_j = (\mu_j; \{u_{j,\tau}\}, \{v_{j,\tau}\})$, $j = 1, 2$.

Let ζ_j be the solutions to trivial control systems $d\zeta_j = d\vartheta_j$, $\zeta_j(t_0) = 0$, $j = 1, 2$. (Thereby, $\zeta_j(t) = F(t; \mu_j)$, where F, as usual, stands for the distribution function of the measure.)

Define the distance in $\mathscr{P}$ by the formula

$$\rho(\wp_1, \wp_2) = \max_{s\in[0,1]} \Big|\mathrm{Ext}[\zeta_1(\cdot), \vartheta_1](s) - \mathrm{Ext}[\zeta_2(\cdot), \vartheta_2](s)\Big| + \int_0^1 \Big|\mathrm{Ext}[u_1(\cdot), \vartheta_1](s) - \mathrm{Ext}[u_2(\cdot), \vartheta_2](s)\Big| ds. \tag{6.4}$$

Functions $\mathrm{Ext}[\zeta_j(\cdot), \vartheta_j]$ are continuous. Functions $\mathrm{Ext}[u_j(\cdot), \vartheta_j]$ are measurable and essentially bounded. Therefore, the maximum and the integral in (6.4) are well defined. It is a straightforward task to prove that $\rho(\cdot, \cdot)$ satisfies all the metric properties, Exercise 6.3. Thus, $(\mathscr{P}, \rho)$ is a metric space. Denote by $\xrightarrow{\rho}$ the convergence of elements in this space. Note that any element (u, v), where $v \in \mathrm{L}_1(T; \mathbb{R}^k)$, $v(t) \in K$ for a.a. t, can be naturally regarded as the element (u, ϑ) of $\mathscr{P}$, where $\vartheta = (\mu; 0; 0)$, and $d\mu = v(t)dt$. Then, notation $\mathrm{Ext}[h, v](\cdot)$ should be read as $\mathrm{Ext}[h, \vartheta](\cdot)$.

The introduced metric satisfies the following two basic properties.

(A) The set of conventional essentially bounded controls (u, v) is everywhere dense in $\mathscr{P}$.

(B) Let $(u_i, \vartheta_i) \xrightarrow{\rho} (u, \vartheta)$, $x_{0,i} \to x_0 \in \mathbb{R}^n$, and $x_i(\cdot)$ be the trajectories corresponding to triples $(x_{0,i}, u_i, \vartheta_i)$ in view of (6.2). If functions $\mathrm{Ext}[u_i, \vartheta_i](s)$, $\mathrm{Ext}[x_i, \vartheta_i](s)$ are uniformly bounded, then there exists trajectory $x(\cdot)$ corresponding to (x_0, u, ϑ) in view of (6.2) such that $\mathrm{Ext}[x_i, \vartheta_i](s) \rightrightarrows \mathrm{Ext}[x, \vartheta](s)$ on $[0, 1]$.

These properties are very similar to those stated in Chap. 4, so they can be derived similarly; see Exercise 6.4. The following lemma deals with ρ-convergence and constitutes an important technical result, needed in the proof of the maximum principle. By $|\cdot|$ in what follows, it is understood the finite-dimensional norm given by the sum of absolute values of the components of a vector.

Lemma 6.1 *Given that* $(\bar{u}_i, \bar{v}_i) \xrightarrow{\rho} (\bar{u}, \bar{\vartheta}) \in \mathscr{P}$ *and*

$$\int_{t_0}^{t_1} \big(|\zeta_i(t) - \bar{\zeta}_i(t)|^2 + |u_i(t) - \bar{u}_i(t)|^2\big) m_i(t)dt \to 0, \tag{6.5}$$

where

$$\zeta_i(t) = \int_{t_0}^{t} v_i d\tau, \qquad \bar{\zeta}_i(t) = \int_{t_0}^{t} \bar{v}_i d\tau, \qquad m_i(t) = 1 + \frac{|v_i(t)| + |\bar{v}_i(t)|}{2},$$

functions u_i, $\bar{u}_i$, v_i, *and* $\bar{v}_i$ *are of class* L_∞, *and functions* u_i *are uniformly bounded.*

Then, $(u_i, v_i) \xrightarrow{\rho} (\bar{u}, \bar{\vartheta})$.

Proof Let us denote by

$$\pi_i(t) = c_i^{-1}\left(t - t_0 + \int_{t_0}^{t} |v_i(\tau)|d\tau\right), \qquad \bar{\pi}_i(t) = (\bar{c}_i)^{-1}\left(t - t_0 + \int_{t_0}^{t} |\bar{v}_i(\tau)|d\tau\right),$$

where $c_i = t_1 - t_0 + \|v_i\|_{L_1}$, $\bar{c}_i = t_1 - t_0 + \|\bar{v}_i\|_{L_1}$. Denote by $\theta_i(s)$, $\bar{\theta}_i(s)$ the corresponding inverse functions.

Consider the following change of variable

$$\pi_i^*(t) = (c_i^*)^{-1} \int_{t_0}^{t} m_i(\tau)d\tau,$$

where $c_i^* = \|m_i\|_{L_1}$, and let θ_i^* be the corresponding inverse function. Let z_i denote $|\zeta_i|$, and $\bar{z}_i = |\bar{\zeta}_i|$.

In view of (6.5), it follows that $z_i(t) \to \bar{z}(t) = F(t, |\bar{\vartheta}|)$ a.a. $t \in T$. Let us establish that $z_i(t_1) \to \bar{z}(t_1)$. This will mean that the variation measures converge in the weak-* topology.

From (6.5), in view of the variable change, we have

$$\int_0^1 |\zeta_i(\theta_i^*(s)) - \bar{\zeta}_i(\theta_i^*(s))|^2 ds \to 0. \tag{6.6}$$

Indeed

$$\zeta_i(\theta_i^*(s)) = \int_{t_0}^{\theta_i^*(s)} v_i(t)dt = \int_{t_0}^{\theta_i^*(s)} \frac{2c_i^* v_i(t)d\pi_i^*(t)}{2 + |v_i(t)| + |\bar{v}_i(t)|} = \int_0^s \beta_i(\tau)d\tau,$$

where it is denoted

$$\beta_i(\tau) = \frac{2c_i^* v_i(\theta_i^*(\tau))}{2 + |v_i(\theta_i^*(\tau))| + |\bar{v}_i(\theta_i^*(\tau))|}.$$

From (6.6), by taking into account $K \subset \mathbb{R}_+^k$,

$$\int_0^1 |z_i(\theta_i^*(s)) - \bar{z}_i(\theta_i^*(s))|^2 ds \to 0.$$

It is also clear that

$$z_i(\theta_i^*(s)) = \int_0^s |\beta_i(\tau)|\, d\tau.$$

Note that $|\beta_i(\tau)| \le 2c_i^*$, and hence, in view of the presented expression, it follows that functions $\zeta_i(\theta_i^*(s))$ and $z_i(\theta_i^*(s))$ are equicontinuous and uniformly bounded.

Then, by the Arzela–Ascoli theorem, after extracting a subsequence, we obtain that $\zeta_i(\theta_i^*(s)) \rightrightarrows \zeta(s)$, $z_i(\theta_i^*(s)) \rightrightarrows z(s)$, where ζ, z are some continuous functions on $[0, 1]$. Analogous reasonings take place for functions $\bar{\zeta}_i(\theta_i^*(s))$, and $\bar{z}_i(\theta_i^*(s))$. In view of (6.6), after extracting a subsequence, we have that $\bar{\zeta}_i(\theta_i^*(s)) \rightrightarrows \zeta(s)$, $\bar{z}_i(\theta_i^*(s)) \rightrightarrows z(s)$. However, $\bar{z}_i(\theta_i^*(1)) = \bar{z}_i(t_1) \to \bar{z}(t_1)$ as $i \to \infty$ due to ρ-convergence. From here, it follows that $z_i(t_1) = z_i(\theta_i^*(1)) \to \bar{z}(t_1)$. Since these arguments can be applied to any subsequence of the original sequence, we have proved that $z_i(t_1) \to \bar{z}(t_1)$.

By virtue of weak-* convergence of the variation measures, it obviously follows that $\bar{\theta}_i(s) \rightrightarrows \bar{\theta}(s)$, $\theta_i(s) \rightrightarrows \bar{\theta}(s)$, and $\theta_i^*(s) \rightrightarrows \bar{\theta}(s)$, where $\bar{\theta}(s)$ is the inverse function to

$$\bar{\pi}(t) = \frac{t - t_0 + \bar{z}(t)}{t_1 - t_0 + \bar{z}(t_1)}.$$

By definition, since $(\bar{u}_i, \bar{v}_i) \xrightarrow{\rho} (\bar{u}, \bar{v})$, we have that $\text{Ext}[\bar{\zeta}_i, \bar{v}_i](s) \rightrightarrows \bar{\zeta}^{\text{ext}}(s) = \text{Ext}[\bar{\zeta}, \bar{\vartheta}](s)$. This means that $\bar{\zeta}_i(\bar{\theta}_i(s)) \rightrightarrows \bar{\zeta}^{\text{ext}}(s)$. Let us show that $\zeta_i(\theta_i(s)) \rightrightarrows \bar{\zeta}^{\text{ext}}(s)$. It is enough to show that there exists a subsequence with such a property.

By definition, we have

$$\frac{\pi_i(\theta_i^*(s)) + \bar{\pi}_i(\theta_i^*(s))}{2} = \pi_i^*(\theta_i^*(s)) = s.$$

Hence, by taking into account that $z_i(\theta_i^*(s)) - \bar{z}_i(\theta_i^*(s)) \rightrightarrows 0$ and the definitions of $\pi_i, \bar{\pi}_i$, it follows that $\pi_i(\theta_i^*(s)) \rightrightarrows s$, $\bar{\pi}_i(\theta_i^*(s)) \rightrightarrows s$.

Note that

$$\zeta_i(\theta_i^*(s)) = \int\limits_0^{\pi_i(\theta_i^*(s))} \frac{c_i v_i(\theta_i(\tau))}{1 + |v_i(\theta_i(\tau))|} d\tau = \text{Ext}[\zeta_i, v_i](\pi_i(\theta_i^*(s))).$$

From this, we obtain $\text{Ext}[\zeta_i, v_i](s) \rightrightarrows \zeta(s)$, since $\zeta_i(\theta_i^*(s)) \rightrightarrows \zeta(s)$, $\pi_i(\theta_i^*(s)) \rightrightarrows s$.

Similarly,

$$\bar{\zeta}_i(\theta_i^*(s)) = \int\limits_0^{\bar{\pi}_i(\theta_i^*(s))} \frac{\bar{c}_i \bar{v}_i(\bar{\theta}_i(\tau))}{1 + |\bar{v}_i(\bar{\theta}_i(\tau))|} d\tau = \text{Ext}[\bar{\zeta}_i, \bar{v}_i](\bar{\pi}_i(\theta_i^*(s))).$$

As mentioned above, $\bar{\zeta}_i(\theta_i^*(s)) \rightrightarrows \zeta(s)$. However, $\text{Ext}[\bar{\zeta}_i, \bar{v}_i](s) \rightrightarrows \bar{\zeta}^{\text{ext}}(s)$; therefore, since $\bar{\pi}_i(\theta_i^*(s)) \rightrightarrows s$, we obtain $\zeta(s) \equiv \bar{\zeta}^{\text{ext}}(s)$. Thus, $\zeta_i(\theta_i(s)) \rightrightarrows \bar{\zeta}^{\text{ext}}(s) \Rightarrow \text{Ext}[\zeta_i, v_i](s) \rightrightarrows \bar{\zeta}^{\text{ext}}(s)$.

In order to prove that $(u_i, v_i) \xrightarrow{\rho} (\bar{u}, \bar{\vartheta})$, it remains to be shown that $\text{Ext}[u_i, v_i](s) \to \bar{u}^{\text{ext}}(s) = \text{Ext}[\bar{u}, \bar{\vartheta}](s)$ in L_1. From (6.5), we have

$$\int\limits_0^1 |u_i(\theta_i^*(s)) - \bar{u}_i(\theta_i^*(s))| ds \to 0. \tag{6.7}$$

In view of ρ-convergence, we have that $\bar{u}_i(\bar{\theta}_i(s)) \to \bar{u}^{\mathrm{ext}}(s)$ in L_1. Denote

$$f_i(s) = \bar{u}_i(\bar{\theta}_i(s)) - \bar{u}^{\mathrm{ext}}(s), \quad \alpha_i(s) = \bar{\pi}_i(\theta_i^*(s)).$$

Note that functions $f_i, (\dot{\alpha}_i)^{-1}$ are uniformly bounded in the L_∞-norm. Here, functions α_i map [0, 1] onto [0, 1], and they are strictly increasing. Therefore,

$$\int_0^1 \big| f_i(\alpha_i(s))\big|\, ds = \int_0^1 \big| f_i(\alpha_i(s))\big| \frac{\sqrt{\dot{\alpha}_i(s)}}{\sqrt{\dot{\alpha}_i(s)}} ds$$

$$\leq \sqrt{\int_0^1 \big| f_i(\alpha_i(s))\big|^2 \dot{\alpha}_i(s)\, ds} \cdot \sqrt{\int_0^1 \frac{1}{\dot{\alpha}_i(s)} ds} \leq \mathrm{const}\sqrt{\int_0^1 \big| f_i(\varsigma)\big|^2 d\varsigma} \to 0.$$

Here, we used the Cauchy inequality and also the fact that if $\| f_i \|_{\mathrm{L}_1} \to 0$ and f_i are uniformly bounded, then $\| f_i \|_{\mathrm{L}_2} \to 0$; see Exercise 6.5. Therefore, it is established that

$$\int_0^1 |\bar{u}_i(\theta_i^*(s)) - \bar{u}^{\mathrm{ext}}(\alpha_i(s))|\, ds \to 0.$$

In view of $\alpha_i(s) \rightrightarrows s$, and Exercise 4.4 of Chap. 4, we have $\bar{u}^{\mathrm{ext}}(\alpha_i(s)) \to \bar{u}^{\mathrm{ext}}(s)$ in L_1. Thus, we have proved that $\bar{u}_i(\theta_i^*(s)) \to \bar{u}^{\mathrm{ext}}(s)$ in L_1. Hence, from (6.7), it follows that $u_i(\theta_i^*(s)) \to \bar{u}^{\mathrm{ext}}(s)$ in L_1. Similarly, though considering $f_i(s) = u_i(\theta_i^*(s)) - \bar{u}^{\mathrm{ext}}(s)$, $\alpha_i(s) = \pi_i^*(\theta_i(s))$, we finally establish that $u_i(\theta_i(s)) \to \bar{u}^{\mathrm{ext}}(s)$ in L_1. For this, we need only to verify whether $\pi_i^*(\theta_i(s)) \rightrightarrows s$.

We have that $\bar{z}_i(\theta_i(\pi_i(\theta_i^*(s)))) = \bar{z}_i(\bar{\theta}_i(\bar{\pi}_i(\theta_i^*(s))))$. By definition $\bar{z}_i(\bar{\theta}_i(s)) \rightrightarrows \bar{c}_i s - \bar{\theta}(s) + t_0$ (in view of $\bar{c}_i \bar{\pi}_i(t) = t - t_0 + \bar{z}_i(t)$). Therefore, since $\bar{\pi}_i(\theta_i^*(s)) \rightrightarrows s$, $\pi_i(\theta_i^*(s)) \rightrightarrows s$, it has been obtained that $\bar{z}_i(\theta_i(s)) \rightrightarrows \bar{c}_i s - \bar{\theta}(s) + t_0$. By definition, we now have

$$\pi_i^*(\theta_i(s)) = \frac{\pi_i(\theta_i(s)) + \bar{\pi}_i(\theta_i(s))}{2} = \frac{s + \bar{\pi}_i(\theta_i(s))}{2} = \frac{\bar{c}_i s + \theta_i(s) - t_0 + \bar{z}_i(\theta_i(s))}{2\bar{c}_i} \rightrightarrows s.$$

The proof is complete. □

Remark 6.1 Note that it was also proved that $\theta_i^* \rightrightarrows \bar{\theta}$, $\zeta_i \circ \theta_i^* \rightrightarrows \bar{\zeta}^{\mathrm{ext}}$ on [0, 1], and $u_i \circ \theta_i^* \to \bar{u}^{\mathrm{ext}}$ in $\mathrm{L}_1([0, 1]; \mathbb{R}^m)$.

Next in this section, bearing in mind the regularity of mixed constraints, we gather several technical results which are essential for the proof of the maximum principle. Consider finite-dimensional Euclidean spaces X and Y. Suppose that $F : X \to Y$ is a smooth map, and $V \subseteq Y$ is a closed set containing zero. Let $x_* \in X$, $F(x_*) = y_*$. For each y, we are interested in the question of existence of a solution x to the inclusion

$$F(x) \in y + V$$

in the neighborhood of (x_*, y_*).

Definition 6.4 The map F is called metrically regular at the point x_* w.r.t. the set V, provided there exist a number $c > 0$ and a neighborhood O of (x_*, y_*) such that:

$$\mathrm{dist}(x, F^{-1}(y + V)) \leq c \cdot \mathrm{dist}(F(x), y + V), \quad \forall\, (x, y) \in O.$$

As usual, we say that the point (x_*, y_*) is regular if

$$\ker F'^*(x_*) \cap N_V(0) = \{0\}.$$

Here, $N_V(0)$ stands for the limiting normal cone to $y_* + V$ at $y = y_*$.

The following result is well known. Its broad generalization within the infinite-dimensional framework can be found, e.g., in [16]. Below, a short proof for the finite-dimensional case is proposed.

Lemma 6.2 *Suppose x_* to be regular. Then, F is metrically regular w.r.t. set V at the point x_*.*

Proof Assume the contrary. Let $F(x)$ not be metrically regular w.r.t. V at x_*. Then, for any δ, and $c > 0$, there exist vectors x_δ and y_δ, which also depend on c such that

$$\mathrm{dist}(x_\delta, F^{-1}(y_\delta + V)) > c \cdot \mathrm{dist}(F(x_\delta), y_\delta + V), \tag{6.8}$$

and $|x_\delta - x_*| \leq \delta$, $|y_\delta - y_*| \leq \delta$.

Let $\xi_\delta \in V$ be such a vector that

$$|F(x_\delta) - y_\delta - \xi_\delta| = \mathrm{dist}(F(x_\delta), y_\delta + V).$$

Denote $\bar{y}_\delta = F(x_\delta) - y_\delta - \xi_\delta$. We have that $\bar{y}_\delta \neq 0$ in view of (6.8).

Consider the extremal problem

$$\begin{aligned} &\text{Minimize} \;\; c^{-1}|x - x_\delta| + t|\bar{y}_\delta|, \\ &\text{subject to} \;\; F(x) - t\bar{y}_\delta \in y_\delta + V, \\ &\qquad\qquad\quad\;\; t \geq 0. \end{aligned} \tag{6.9}$$

A solution to (6.9) does exist due to the form of the cost function and the fact that there is at least one admissible point satisfying the constraints (take $x = x_\delta$, $t = 1$). Denote this solution by $\hat{x}_\delta$, $\hat{t}_\delta$. Note that

$$\hat{t}_\delta > 0. \tag{6.10}$$

Indeed, otherwise, we would have

$$F(\hat{x}_\delta) \in y_\delta + V \;\; \Rightarrow \;\; |\hat{x}_\delta - x_\delta| \geq \mathrm{dist}(x_\delta, F^{-1}(y_\delta + V)).$$

However, in view of (6.8), it holds that

$$\mathrm{dist}(x_\delta, F^{-1}(y_\delta + V)) > c \cdot \mathrm{dist}(F(x_\delta), y_\delta + V) = c|\bar{y}_\delta|.$$

Therefore, the minimal value of the cost function in (6.9) is greater than $|\bar{y}_\delta|$. However, this is not possible as the value of the cost is precisely $|\bar{y}_\delta|$ at $x = x_\delta, t = 1$. Thereby, (6.10) holds true.

Next, we apply the Lagrange multiplier rule from [17] (Theorem 5.5) to problem (6.9). There exist a number $\lambda_\delta^0 \geq 0$, a vector $\lambda_\delta \in N_V(\zeta_\delta)$, where $\zeta_\delta = F(\hat{x}_\delta) - \hat{t}_\delta \bar{y}_\delta - y_\delta$, and a vector $b_\delta \in B_X$, such that

$$\lambda_\delta^0 c^{-1} b_\delta = -\lambda_\delta F'(\hat{x}_\delta), \tag{6.11}$$

$$\lambda_\delta^0 |\bar{y}_\delta| = \langle \lambda_\delta, \bar{y}_\delta \rangle, \tag{6.12}$$

$$\lambda_\delta^0 + |\lambda_\delta| = 1. \tag{6.13}$$

Here, in (6.12), Condition (6.10) has been used. Since $\bar{y}_\delta \neq 0$, it can be deduced from (6.12) that $\lambda_\delta^0 \leq |\lambda_\delta|$. However, in view of (6.13), we have $\lambda_\delta^0 = 1 - |\lambda_\delta|$ and, therefore,

$$|\lambda_\delta| \geq \frac{1}{2}. \tag{6.14}$$

Now, pass consecutively to the limit, first as $\delta \to 0$ and afterward as $c \to \infty$. By extracting a subsequence, in view of (6.13) and (6.14), we conclude that there exists a nonzero vector $\lambda \in N_V(0)$, such that $\lambda_\delta \to \lambda$. By passing to the limit in (6.11) as $\delta \to 0$, $c \to \infty$, we obtain $0 = \lambda F'(x_*)$, which is contradictory to the Robinson qualification condition as point x_* is assumed to be regular. This contradiction proves that the map F is metrically regular. □

From the proof, an even a more precise statement follows.

Denote

$$\psi(x, y) = \mathrm{sur} M(x|y),$$

where $M(x) = F(x) - V$, $y \in M(x)$. Consider a number $\alpha \in (0, L)$, where $L := \psi(x_*, y_*)$. Consider the scalar function

$$\omega(\delta) = \min_{x:\,|x-x_*|\leq r_x(\delta)} \; \min_{y:\,|y-y_*|\leq r_y(\delta)} \psi(x, y),$$

where

$$r_x(\delta) = \delta + \tfrac{\delta + p(\delta)}{\alpha},$$
$$r_y(\delta) = 2\delta + p(\delta),$$
$$p(\delta) = \max_{x:\,|x-x_*|\leq\delta} |F(x) - F(x_*)|.$$

Obviously, function $\omega(\delta)$ is decreasing when $\delta > 0$ and continuous at $\delta = 0$ with $\omega(0) = L$. This is due to the fact that the modulus of surjection is a lower semicontinuous function.

Then, for a given $\delta > 0$, such that $\omega(\delta) > \alpha$, the constant $c > 0$ and the set O can be chosen such that,

$$c \cdot \alpha = 1, \tag{6.15}$$

$$O = (x_*, y_*) + \delta \cdot B_{X \times Y}. \tag{6.16}$$

Indeed, by assuming the opposite, we obtain x_δ and y_δ and then, by using the same construction as in the proof of Lemma 6.2, we come to a contradiction. Due to the fact that $|\bar{y}_\delta| \leq p(\delta) + \delta$, it is simple to also extract the following estimates from the proof,

$$|\hat{x}_\delta - x_*| \leq r_x(\delta), \quad |\hat{y}_\delta - y_*| \leq r_y(\delta), \quad \text{where } \hat{y}_\delta := \hat{t}_\delta \bar{y}_\delta + y_\delta.$$

Due to $\lambda_\delta^0 \leq |\lambda_\delta|$, from (6.11), we have

$$|\lambda_\delta| \geq c \cdot |\lambda_\delta F'(\hat{x}_\delta)|.$$

Therefore, after setting $e_\delta = \lambda_\delta |\lambda_\delta|^{-1}$,

$$1 \geq c \cdot |e_\delta F'(\hat{x}_\delta)| \geq c \cdot \psi(\hat{x}_\delta, \hat{y}_\delta)|e_\delta| \geq c \cdot \omega(\delta) > c \cdot \alpha = 1.$$

This contradiction leads to the following assertion.

Lemma 6.3 *Suppose that x_* is regular. Let $\alpha \in (0, L)$. Then, F is metrically regular w.r.t. the set V at the point x_* with $c = \alpha^{-1}$ and $O = (x_*, y_*) + \delta \cdot B_{X \times Y}$, where δ is an arbitrary number such that $\omega(\delta) > \alpha$.*[2]

Returning to the mixed constraint system $r(x, u, t) \in C$, take a point $t \in T$. In view of the obvious relation $\ker \frac{\partial r^*}{\partial(x,u)}(x, u, t) \subseteq \ker \frac{\partial r}{\partial u}^*(x, u, t)$, and Lemma 6.2, function $r(\cdot, t)$, as function of x, u, is metrically regular w.r.t. $C - r(x, u, t)$ at any point (x, u), where $u \in U(x, t)$ is regular.

Let a conventional trajectory $x(\cdot)$ be regular w.r.t. the mixed constraints and sets $U(x(t), t)$ be uniformly bounded. From the properties of r, it follows the existence of a number ν, which does not depend on t, such that, for all t,

$$\text{sur} M(x, u, t) \geq \varepsilon_0/2 \ \ \forall u \in U(x(t), t) + \nu B_{\mathbb{R}^m}, \ \forall x \in x(t) + \nu B_{\mathbb{R}^n}.$$

Let $u(t) \in U(x(t), t)$ be any measurable selector and $\{x_i(t)\}$ be a sequence of conventional trajectories converging uniformly to $x(t)$. Then, in view of metric regularity, applied for $t \in T$, with $y = 0$, $c = \frac{3}{\varepsilon_0}$, and by taking into account the above estimate, conditions (6.15), (6.16), and also the definition of $p(\delta)$, $r_x(\delta)$, $r_y(\delta)$, and

[2] When $\alpha \to L$, the number c approaches the number $\text{sur} M(x_*|y_*)^{-1}$ which is called the modulus of regularity and designated by $\text{reg} M(x_*|y_*)$. It follows that the modulus of regularity is the lower bound of all such c, for which the estimate of metric regularity still holds true; see [16].

$\omega(\delta)$ and, again, the continuity of r, we derive that there is a sufficiently large number N such that, for all $i \geq N$, the following inequality is valid

$$\text{dist}(u(t), U(x_i(t), t)) \leq c\text{dist}(r(x_i(t), u(t), t), C) \text{ a.a. } t.$$

The right-hand side of this inequality converges uniformly to zero. Then, by using the measurable selection lemma (Filippov's lemma, [11]), the following assertion can be obtained.

Lemma 6.4 *Let $x_i \in C(T; \mathbb{R}^n)$, $x_i \rightrightarrows x$, and $u(t) \in U(x(t), t)$ for a.a. t, where $x(\cdot)$ is regular w.r.t. the mixed constraints, while sets $U(x(t), t)$ are uniformly bounded w.r.t. $t \in T$. Then, for each sufficiently large i, there exists a measurable function $u_i(\cdot)$ such that $u_i(t) \in U(x_i(t), t)$ for a.a. t, and*

$$\|u_i - u\|_{L_\infty} \to 0.$$

This property constitutes the earlier mentioned uniform w.r.t. t continuity of the map $U(x, t)$ w.r.t. x-variable. It will be used in the next section in the proof of the maximum principle.

6.5 Maximum Principle

Denote by H the Hamilton–Pontryagin function

$$H(x, u, \psi, t) := \langle f(x, u, t), \psi \rangle,$$

and by Q the vector-valued function $Q(x, u, \psi, t) := G^*(x, u, t)\psi$.

Theorem 6.1 *Let $(\hat{x}, \hat{u}, \hat{\vartheta})$, $\hat{\vartheta} = (\hat{\mu}; \{\hat{u}_\tau\}, \{\hat{v}_\tau\})$ be an optimal process in Problem (6.1), while the extended trajectory $(\hat{x}; \{\hat{x}_\tau\})$ is regular w.r.t. the mixed constraints. Let also the sets $U(t)$, $U_\tau(s)$ be uniformly bounded for all $t \in T$, $s \in [0, 1]$, $\tau \in Ds(\hat{\mu})$.*

Then, there exist a number $\lambda \geq 0$, a vector-valued function of bounded variation ψ, a measurable integrable vector-valued function η, $\eta(t) \in \text{conv}N_C(r(t))$ a.a. t, both defined on T, and, for every point $\tau \in Ds(\hat{\mu})$, there exist an absolutely continuous vector-valued function ψ_τ and an essentially bounded vector-valued function η_τ, $\eta_\tau(s) \in \text{conv}N_C(r_\tau(s))$ a.a. s, both defined on the interval $[0, 1]$, such that

$$\lambda + |\psi(t)| \neq 0, \ \forall t \in T, \ \ \lambda + |\psi_\tau(s)| \neq 0, \ \forall s \in [0, 1], \ \forall \tau \in Ds(\hat{\mu}), \quad (6.17)$$

$$\psi(t) = \psi(t_0) - \int_{t_0}^{t} \frac{\partial H}{\partial x}(\varsigma)d\varsigma - \int_{[t_0,t]} \frac{\partial}{\partial x}\Big\langle Q(\varsigma), d\hat{\mu}_{\mathrm{c}}\Big\rangle$$
$$+ \int_{t_0}^{t} \frac{\partial r^*}{\partial x}(\varsigma)\eta(\varsigma)d\varsigma + \sum_{\tau \in Ds(\hat{\mu}):\, \tau \le t} \Big(\psi_\tau(1) - \psi(\tau^-)\Big), \quad \forall t \in (t_0, t_1], \tag{6.18}$$

$$\begin{cases} \dot{\hat{x}}_\tau(s) = G_\tau(s)\hat{v}_\tau(s), \\ \dot{\psi}_\tau(s) = -\frac{\partial}{\partial x}\big\langle Q_\tau(s), \hat{v}_\tau(s)\big\rangle + \frac{\partial r_\tau^*}{\partial x}(s)\eta_\tau(s), \\ \hat{x}_\tau(0) = \hat{x}(\tau^-), \ \ \psi_\tau(0) = \psi(\tau^-), \\ s \in [0, 1], \ \forall \tau \in Ds(\hat{\mu}), \end{cases} \tag{6.19}$$

$$(\psi(t_0), -\psi(t_1)) \in \lambda\varphi'(\hat{p}) + N_S(\hat{p}), \tag{6.20}$$

$$\max_{u \in U(t)} H(u, t) = H(t), \quad a.a.\, t \in T, \tag{6.21}$$

$$\begin{cases} \max\limits_{v \in K} \max\limits_{u \in U(t)} \langle Q(u, t), v\rangle = 0 \ \ \forall t \in T, \\ \max\limits_{v \in K} \max\limits_{u \in U_\tau(s)} \langle Q_\tau(u, s), v\rangle = \\ = \big\langle Q_\tau(s), \hat{v}_\tau(s)\big\rangle = 0, \ \ a.a.\, s \in [0, 1] \ \forall \tau \in Ds(\hat{\mu}), \\ \int_{[t_0,t]} \Big\langle Q(\varsigma), d\hat{\mu}_{\mathrm{c}}\Big\rangle = 0 \ \ \forall t \in T, \end{cases} \tag{6.22}$$

$$\begin{cases} \dfrac{\partial H}{\partial u}(t) + \dfrac{\partial}{\partial u}\Big\langle Q(t), \dfrac{d\hat{\mu}_{\mathrm{ac}}}{dt}\Big\rangle = \dfrac{\partial r^*}{\partial u}(t)\eta(t), \ \ a.a.\, t \in T, \\ \int_{[t_0,t]} \frac{\partial}{\partial u}\Big\langle Q(\varsigma), d\hat{\mu}_{\mathrm{sc}}\Big\rangle = 0 \ \ \forall t \in T, \\ \dfrac{\partial}{\partial u}\Big\langle Q_\tau(s), \hat{v}_\tau(s)\Big\rangle = \dfrac{\partial r_\tau^*}{\partial u}(s)\eta_\tau(s), \ \ a.a.\, s \in [0, 1], \ \ \forall \tau \in Ds(\hat{\mu}), \end{cases} \tag{6.23}$$

where $\hat{\mu}_{\mathrm{ac}}$, $\hat{\mu}_{\mathrm{sc}}$ *are the absolutely continuous and singular continuous components of the measure* $\hat{\mu}$, *respectively:* $\hat{\mu}_{\mathrm{c}} = \hat{\mu}_{\mathrm{ac}} + \hat{\mu}_{\mathrm{sc}}$.

Here, as usual, if some arguments of a function or a set-valued mapping of x, u, ψ, t are omitted, then it means that the extremal values $\hat{x}(t)$, $\hat{u}(t)$, $\psi(t)$ substitute the omitted arguments everywhere. For example, $H(u, t) = H(\hat{x}(t), u, \psi(t), t)$, or $U(t) = U(\hat{x}(t), t)$. Similarly, the subscript τ at some function or a set-valued mapping means that the attached values $\hat{x}_\tau(s)$, $\hat{u}_\tau(s)$, $\psi_\tau(s)$ at the point of the discontinuity τ substitute the omitted arguments. For example, $r_\tau(s) = r(\hat{x}_\tau(s), \hat{u}_\tau(s), \tau)$. The same notation is used for the partial derivatives of functions w.r.t. x, u.

Proof Let the triple $(\hat{x}, \hat{u}, \hat{\vartheta})$ be optimal for the problem (6.1). It is not restrictive to assume that $\varphi(\hat{p}) = 0$ and also that $f(\cdot, t) = 0$ for all points t in the support of the singular component of the measure $|\hat{\vartheta}|$. Indeed, the support of the singular component has zero Lebesgue measure. Therefore, the values of f on this set do not affect the trajectory evolution. Moreover, now, $\hat{f}(t)$ is measurable w.r.t. $\ell + |\hat{\vartheta}|$.

Let us take positive numbers $\bar{c}$, and δ such that

$$|u| \leq \bar{c}, \quad \forall\, u \in U(x,t), \ \forall\, x \text{ s.t. } |x - \hat{x}(t)| \leq \delta, \ \forall\, t \in T, \text{ and}$$

$$\forall\, u \in U(z,\tau), \ \forall\, z \text{ s.t. } |z - \hat{x}_\tau(s)| \leq \delta, \ \forall\, s \in [0,1], \ \forall\, \tau \in \mathrm{Ds}(\hat{\mu}).$$

This is possible due to the uniform boundedness of the sets $U(t)$, $U_\tau(t)$ and the properties of map r. Moreover, number δ can be chosen such that, for some sufficiently small $\theta > 0$, the following estimate holds true for all $t \in T$, $s \in [0,1]$, and $\tau \in \mathrm{Ds}(\hat{\mu})$

$$\begin{cases} 2\left|y\frac{\partial r}{\partial u}(x,u,t)\right| \geq \varepsilon_0 |y|, \ \forall\, y \in N_C(\xi), \ \forall\, \xi \in \Pi_C(r(x,u,t)) \\ \forall\, (x,u) \text{ s.t. } |x - \hat{x}(t)| \leq \delta, \ |u| \leq c, \ \mathrm{dist}(r(x,u,t), C) \leq \theta, \\ 2\left|y\frac{\partial r}{\partial u}(x,u,\tau)\right| \geq \varepsilon_0 |y|, \ \forall\, y \in N_C(\xi), \ \forall\, \xi \in \Pi_C(r(x,u,\tau)) \\ \forall\, (x,u) \text{ s.t. } |x - \hat{x}_\tau(s)| \leq \delta, \ |u| \leq c, \ \mathrm{dist}(r(x,u,\tau), C) \leq \theta. \end{cases} \tag{6.24}$$

Here, number ε_0 comes from the definition of regularity (see Definition 6.3), Π_C designates the Euclidean projection onto the set C, and c is any number greater than $\bar{c}$. Estimate (6.24) follows from the regularity of the optimal arc, the upper semicontinuity of the limiting normal cone $N_C(y)$ w.r.t. y, and from the continuity properties of r.

Consider the discontinuous time variable change $\hat{\pi}(t)$ associated with the optimal impulsive control $\hat{\vartheta}$. The inverse function is denoted by $\hat{\theta}(s) : [0,1] \to T$.

Consider functions

$$\hat{m}_1(t) = \frac{a \cdot dt}{d(t + |\hat{\vartheta}|)}, \quad \hat{m}_2(t) = \frac{a \cdot d\hat{\mu}_c}{d(t + |\hat{\vartheta}|)}, \quad \text{where } a = t_1 - t_0 + |\hat{\vartheta}|(T),$$

which are the calibrated Radon–Nikodym derivatives of measures ℓ and $\hat{\mu}_c$ w.r.t. Lebesgue–Stieltjes measure $\ell + |\hat{\vartheta}|$, respectively. They are $(\ell + |\hat{\vartheta}|)$-measurable and bounded.

Denote $\hat{\Gamma}_\tau = [\hat{\pi}(\tau^-), \hat{\pi}(\tau^+)]$, $\tau \in \mathrm{Ds}(\hat{\mu})$, and define

$$\hat{\alpha}(\varsigma) = \begin{cases} \hat{m}_1(\hat{\theta}(\varsigma)), & \text{if } \varsigma \notin \bigcup_{\tau \in \mathrm{Ds}(\hat{\mu})} \hat{\Gamma}_\tau, \\ 0, & \text{otherwise,} \end{cases}$$

$$\hat{\beta}(\varsigma) = \begin{cases} \ell^{-1}(\hat{\Gamma}_\tau)\hat{v}_\tau(\hat{\xi}_\tau(\varsigma)), & \text{if } \exists\, \tau \in \mathrm{Ds}(\hat{\mu}) : \ \varsigma \in \hat{\Gamma}_\tau, \\ \hat{m}_2(\hat{\theta}(\varsigma)), & \text{otherwise,} \end{cases}$$

where $\hat{\xi}_\tau(\varsigma)$ is the calibration function associated with the optimal impulsive control.

Let us note that (6.17)–(6.23) are equivalent to the following conditions:

$$\lambda + |\psi^{\text{ext}}(s)| \neq 0 \ \ \forall s \in [0, 1], \tag{6.25}$$

$$\begin{aligned}\psi^{\text{ext}}(s) = \psi^{\text{ext}}(0) - \int_0^s \frac{\partial H^{\text{ext}}}{\partial x}(\varsigma)\hat{\alpha}(\varsigma)d\varsigma \\ - \int_0^s \frac{\partial}{\partial x}\Big\langle Q^{\text{ext}}(\varsigma), \hat{\beta}(\varsigma)\Big\rangle d\varsigma + \int_0^s \frac{\partial r^{\text{ext}*}}{\partial x}(\varsigma)\eta^{\text{ext}}(\varsigma)d\varsigma \ \ \forall s \in [0, 1],\end{aligned} \tag{6.26}$$

$$(\psi^{\text{ext}}(0), -\psi^{\text{ext}}(1)) \in \lambda\varphi'(\hat{p}) + N_S(\hat{p}), \tag{6.27}$$

$$\max_{u \in U^{\text{ext}}(s)} \hat{\alpha}(s) H(\hat{x}^{\text{ext}}(s), u, \psi^{\text{ext}}(s), \hat{\theta}(s)) = \hat{\alpha}(s) H^{\text{ext}}(s) \ \text{ a.a. } s, \tag{6.28}$$

$$\begin{aligned}\max_{v \in K} \max_{u \in U^{\text{ext}}(s)} \Big\langle Q(\hat{x}^{\text{ext}}(s), u, \psi^{\text{ext}}(s), \hat{\theta}(s)), v\Big\rangle = \\ = \Big\langle Q^{\text{ext}}(s), \hat{\beta}(s)\Big\rangle \ \text{a.a. } s,\end{aligned} \tag{6.29}$$

$$\begin{aligned}\frac{\partial H^{\text{ext}}}{\partial u}(s)\hat{\alpha}(s) + \frac{\partial}{\partial u}\Big\langle Q^{\text{ext}}(s), \hat{\beta}(s)\Big\rangle \\ = \frac{\partial r^{\text{ext}*}}{\partial u}(s)\eta^{\text{ext}}(s) \ \text{a.a. } s \in [0, 1],\end{aligned} \tag{6.30}$$

where the extension is considered w.r.t. $\hat{\vartheta}$, and

$$\psi^{\text{ext}}(s) = \begin{cases} \psi_\tau(\hat{\xi}_\tau(s)), & \text{if } \exists\, \tau \in \text{Ds}(\hat{\mu}) : \ s \in \hat{\Gamma}_\tau, \\ \psi(\hat{\theta}(s)), & \text{otherwise,} \end{cases}$$

$$\eta^{\text{ext}}(s) = \begin{cases} \ell^{-1}(\hat{\Gamma}_\tau)\eta_\tau(\hat{\xi}_\tau(s)), & \text{if } \exists\, \tau \in \text{Ds}(\hat{\mu}) : \ s \in \hat{\Gamma}_\tau, \\ \hat{\alpha}(s)\eta(\hat{\theta}(s)), & \text{otherwise.} \end{cases}$$

This is due to the discontinuous time variable change (details are omitted here; see Chap. 2). Recall that, e.g., notation $H^{\text{ext}}(s)$ stands for $H(\hat{x}^{\text{ext}}(s), u, \psi^{\text{ext}}(s), \hat{\theta}(s))$, while other mappings and their derivatives are treated similarly.

In order to prove the theorem, it is enough to derive conditions (6.25)–(6.29) and to establish that $\eta^{\text{ext}} \in \text{L}_\infty([0, 1]; \mathbb{R}^d)$. Indeed, if η^{ext} is bounded, then function $\eta(t) = \frac{\eta^{\text{ext}}(\hat{\pi}(t))}{\hat{m}_1(t)}$ is integrable, because $\hat{m}_1(t)^{-1}$ is integrable; see Exercise 6.6.

Let us proceed with the details. By taking into account property (A) stated above, there exists a sequence of nonimpulsive controls $\bar{v}_i(t) \in \text{L}_\infty(T; \mathbb{R}^k)$, such that $\bar{v}_i(t) \in K$ a.a. t, $\bar{v}_i \xrightarrow{w} |\hat{\vartheta}|$, and $\text{Ext}[\bar{\zeta}_i, \bar{v}_i](s) \rightrightarrows \hat{\zeta}^{\text{ext}}(s) = \text{Ext}[\hat{\zeta}, \hat{\vartheta}](s)$, where $d\bar{v}_i/dt = |\bar{v}_i(t)|$ and $\bar{\zeta}_i, \hat{\zeta}$ are the solutions of the system $d\zeta = d\vartheta$ corresponding to $\bar{v}_i, \hat{\vartheta}$. Thus, we have

$$\int_{t_0}^{\bar{\theta}_i(s)} \bar{v}_i(\tau)d\tau \rightrightarrows \hat{\zeta}^{\text{ext}}(s),$$

where $\bar{\theta}_i(s)$ is the inverse function of

$$\bar{\pi}_i(t) = \Big(t - t_0 + \int_{t_0}^{t} |\bar{v}_i(\tau)|d\tau\Big) \cdot \Big(t_1 - t_0 + \int_{t_0}^{t_1} |\bar{v}_i(\tau)|d\tau\Big)^{-1}.$$

It is simple matter to verify that $\bar{\theta}_i(s) \rightrightarrows \hat{\theta}(s)$, where $\hat{\theta}$ is the inverse function of the discontinuous time variable change associated with $\hat{\vartheta}$. Consider the function $\hat{u}^{\text{ext}}(s) = \text{Ext}[\hat{u}, \hat{\vartheta}](s)$. Set $\bar{u}_i(t) = \hat{u}^{\text{ext}}(\bar{\pi}_i(t))$. By using Exercise 4.4 of Chap. 4, $(\bar{u}_i, \bar{v}_i) \xrightarrow{\rho} (\hat{u}, \hat{\vartheta})$. From here, by using standard arguments, the properties of f, G, and property (B), we derive that trajectories $\bar{x}_i(t)$ corresponding to $(\bar{u}_i, \bar{v}_i)$ (in view of the dynamics of the problem) exist as i is sufficiently large and their sequence converges to $\hat{x}(t)$ as $i \to \infty$, albeit in the extended sense, i.e., $\bar{x}_i(\bar{\theta}_i(s)) \rightrightarrows \hat{x}^{\text{ext}}(s) = \text{Ext}[\hat{x}, \hat{\vartheta}](s)$.

Note that $\bar{u}_i(t) \in cB_{\mathbb{R}^m}$, a.a. t. Take $\kappa_i = i + \text{ess sup}_{t\in T} |\bar{v}_i(t)|^2$. Denote by $\mathcal{M} \subseteq \mathbb{R}^n \times L_1(T; \mathbb{R}^m) \times L_1(T; \mathbb{R}^k)$ the set of the triples $(x_0, u(\cdot), v(\cdot))$ such that $u(t) \in cB_{\mathbb{R}^m}$ a.a. $t \in T$, $v(t) \in K \cap \kappa_i B_{\mathbb{R}^k}$, and there exists on T trajectory $x(t)$ such that $x(t_0) = x_0$, $\dot{x}(t) = f(x(t), u(t), t) + G(x(t), u(t), t)v(t)$, a.a. $t \in T$, and inequality $|x(t)| \le \|\hat{x}^{\text{ext}}\|_C + 1$ $\forall t \in T$ is satisfied. Note that $(\hat{x}_0, \bar{u}_i(\cdot), \bar{v}_i(\cdot)) \in \mathcal{M}$ $\forall i$. The set $\mathcal{M}$ is closed, and, hence, it is a complete metric space w.r.t. the metric generated by norm $|x_0| + \|u\|_{L_1} + \|v\|_{L_1}$.

Let $\varepsilon > 0$ be a positive number. Let $\varphi_\varepsilon(p) = (\varphi(p) + \varepsilon^2/2)^+$, where $a^+ = \max\{a, 0\}$. Let α, β be nonnegative numbers. Let

$$\Delta(\alpha, \beta) = \begin{cases} \alpha\beta^{-2}, \ \beta > 0, \\ 1, \ \alpha > 0, \ \beta = 0, \\ 0, \ \alpha = \beta = 0. \end{cases}$$

Define the functional $D_i(u(\cdot), v(\cdot)) = \displaystyle\int_{t_0}^{t_1} V_i(\zeta(t), u(t), v(t), t)dt$, where

$$V_i(\zeta, u, v, t) = \Big(|\zeta - \bar{\zeta}_i(t)|^2 + |u - \bar{u}_i(t)|^2\Big)\Big(1 + \frac{|v| + |\bar{v}_i(t)|}{2}\Big),$$

$$\zeta(t) = \int_{t_0}^{t} v(\tau)d\tau, \quad \bar{\zeta}_i(t) = \int_{t_0}^{t} \bar{v}_i(\tau)d\tau.$$

Consider the following functional on space $\mathcal{M}$

$$\Phi_i(x_0, u(\cdot), v(\cdot), \varepsilon) = \varphi_\varepsilon(p) + \varepsilon D_i(u(\cdot), v(\cdot)) + \Delta\Big((\text{dist}(p, S))^2, \varphi_\varepsilon(p)\Big) +$$

$$\Delta\Big(\int_{t_0}^{t_1} (\text{dist}(r(x(t), u(t), t), C))^2\Big(1 + \frac{|v(t)| + |\bar{v}_i(t)|}{2}\Big)dt, \varphi_\varepsilon(p)\Big).$$

Here, $p = (x(t_0), x(t_1))$ and $x(\cdot)$ satisfies the equation

$$\dot{x}(t) = f(x(t), u(t), t) + G(x(t), u(t), t)v(t), \quad t \in T.$$

Functional Φ_i is nonnegative and lower semicontinuous on $\mathcal{M}$. This follows directly from the definition. Note that for every fixed ε, we have $\Phi_i(\hat{x}_0, \bar{u}_i, \bar{v}_i, \varepsilon) \to \varepsilon^2/2$ as $i \to \infty$. Then, it is not restrictive to assume that $\Phi_i(\hat{x}_0, \bar{u}_i, \bar{v}_i, \varepsilon_i) \le \varepsilon_i^2$, where $\varepsilon_i := 1/i \to 0+$, since this can always be achieved by extracting a subsequence.

Consider the following problem

$$\Phi_i(x_0, u(\cdot), v(\cdot), \varepsilon_i) \to \min, \quad (x_0, u, v) \in \mathcal{M}.$$

Let us apply Ekeland's variational principle, [10], to this problem at the point $(\hat{x}_0, \bar{u}_i, \bar{v}_i)$. Then, for every i, there exists an element $(x_{0,i}, u_i, v_i) \in \mathcal{M}$ such that

$$\Phi_i(x_{0,i}, u_i, v_i, \varepsilon_i) \le \Phi_i(\hat{x}_0, \bar{u}_i, \bar{v}_i, \varepsilon_i) \le \varepsilon_i^2, \tag{6.31}$$

$$|x_{0,i} - \hat{x}_0| + \int_{t_0}^{t_1} |u_i(t) - \bar{u}_i(t)|dt + \int_{t_0}^{t_1} |v_i(t) - \bar{v}_i(t)|dt \le \varepsilon_i, \tag{6.32}$$

while the triple $(x_{0,i}, u_i(\cdot), v_i(\cdot))$ is the solution to the problem

$$\begin{aligned}
&\text{Minimize } \varphi_{\varepsilon_i}(p) + z_0^{-2}(\text{dist}(p, S))^2 + \\
&\qquad \int_{t_0}^{t_1} z^{-2}(\text{dist}(r(x, u, t), C))^2\Big(1 + \frac{|v| + |\bar{v}_i(t)|}{2}\Big)dt + \\
&\qquad \varepsilon_i\Big(|x_0 - x_{0,i}| + \int_{t_0}^{t_1}\Big(|u - u_i(t)| + |v - v_i(t)| + V_i(\zeta, u, v, t)\Big)dt\Big) \\
&\text{subject to: } \dot{x} = f(x, u, t) + G(x, u, t)v, \\
&\qquad \dot{\zeta} = v, \ \zeta_0 = 0, \\
&\qquad \dot{z} = 0, \ z_0 = \varphi_{\varepsilon_i}(p), \\
&\qquad z_0 > 0, \ |x(t)| \le \|\hat{x}^{\text{ext}}\|_C + 1 \ \forall t \in T, \\
&\qquad u(t) \in cB_{\mathbb{R}^m}, \ v(t) \in K, \ |v(t)| \le \kappa_i \text{ a.a. } t \in T.
\end{aligned} \tag{6.33}$$

Let us explain why it is already assumed above that $\varphi_{\varepsilon_i}(p_i) > 0$, where $p_i = (x_{0,i}, x_{1,i})$, is the optimal pair, and why, respectively, z_0 is considered greater than zero. Indeed, if $\varphi_{\varepsilon_i}(p_i) = 0$, then either the mixed or the endpoint constraints are violated in Problem (6.1). Let the endpoint constraints be violated. Then, it follows that

$$\Delta\Big((\mathrm{dist}(p_i, S))^2, \varphi_{\varepsilon_i}(p_i)\Big) = \Delta\Big((\mathrm{dist}(p_i, S))^2, 0\Big) = 1,$$

which, in view of (6.31), is not possible as $i > 1$. So, $\varphi_{\varepsilon_i}(p_i) > 0$, and therefore, the extra constraint $z_0 > 0$ is not restrictive.

Let us denote the optimal trajectories in problem (6.33) as x_i, ζ_i, z_i. From (6.31), we extract that $\int_{t_0}^{t_1} V_i(t)dt \to 0$ as $i \to \infty$. By using this fact, from Lemma 6.1, it follows that $(u_i, v_i) \xrightarrow{\rho} (\hat{u}, \hat{\vartheta})$. From (6.32), and by extracting a subsequence, we have that $x_{0,i} \to \hat{x}_0$. Then, the application of property (B) gives $\mathrm{Ext}[x_i, v_i] \rightrightarrows \mathrm{Ext}[\hat{x}, \hat{\vartheta}]$. Thus, the solutions to auxiliary i-problems approximate the optimal process.

Note that problem (6.33) is conventional, as it does not comprise impulsive controls. Moreover, it has neither endpoint nor mixed constraints, while the extra state constraint $|x(t)| \le \|\hat{x}^{\mathrm{ext}}\|_C + 1$ is not active as i large due to the above considerations. Thereby, Problem (6.33) is the so-called simplest problem. Then, for every i, there exist a number $\lambda_i > 0$ and absolutely continuous conjugate functions ψ_i^x, ψ_i^z, and ψ_i^ζ such that the maximum principle derived in [17], Theorem 6.27, is true. Consideration of the conditions of this maximum principle suggests the following change of variable:

$$\pi_i^*(t) = \Big(\int_{t_0}^{t_1} m_i(\tau)d\tau\Big)^{-1} \int_{t_0}^{t} m_i(\tau)d\tau,$$

where $m_i(t) = 1 + (|v_i(t)| + |\bar{v}_i(t)|)/2$. After this variable change, the conditions of the maximum principle for i-problem take the form:

$$\begin{aligned}
\psi_i^{\mathrm{ext}}(s) &= \psi_i^{\mathrm{ext}}(0) - \int_0^s \frac{\partial H_i}{\partial x}(\psi_i^{\mathrm{ext}}(\varsigma), \theta_i^*(\varsigma))\alpha_i(\varsigma)d\varsigma - \\
&\int_0^s \tfrac{\partial}{\partial x}\Big\langle Q_i(\psi_i^{\mathrm{ext}}(\varsigma), \theta_i^*(\varsigma)), \beta_i(\varsigma)\Big\rangle d\varsigma + \int_0^s \tfrac{\partial r_i^*}{\partial x}(\theta_i^*(\varsigma))\eta_i^{\mathrm{ext}}(\varsigma)d\varsigma, \\
\xi_i(s) &= \int_s^1 2\lambda_i z_i^{-3}(\mathrm{dist}(r_i(\theta_i^*(\varsigma))), C))^2 d\varsigma, \\
\sigma_i(s) &= \int_s^1 2\lambda_i\varepsilon_i(\zeta_i^{\mathrm{ext}} - \bar{\zeta}_i^{\mathrm{ext}})\alpha_i(\varsigma)d\varsigma, \quad s \in [0, 1],
\end{aligned} \tag{6.34}$$

$$\begin{aligned}
(\psi_i^{\mathrm{ext}}(0), -\psi_i^{\mathrm{ext}}(1)) \in \Big(\lambda_i - \xi_i(0) - \rho_i z_{0,i}^{-1}\mathrm{dist}(p_i, S)\Big)\varphi'(p_i) \\
+ \rho_i \partial\mathrm{dist}(p_i, S) + (\lambda_i\varepsilon_i B_{\mathbb{R}^n}, 0),
\end{aligned} \tag{6.35}$$

$$\begin{cases} \max\limits_{v\in K\cap\kappa_i B_{\mathbb{R}^k}} \max\limits_{u\in cB_{\mathbb{R}^m}} \alpha_i(s)\Big\{H_i(u,\psi_i^{\text{ext}}(s),\theta_i^*(s)) + \\ \big\langle Q_i(u,\psi_i^{\text{ext}}(s),\theta_i^*(s)),v\big\rangle + \big\langle\sigma_i(s),v\big\rangle - \\ \lambda_i z_i^{-2}(\text{dist}(r_i(u,\theta_i^*(s)),C))^2\Big(1+\frac{|v|+|\bar{v}_i(\theta_i^*(s))|}{2}\Big) - \\ \lambda_i\varepsilon_i\Big(|u-u_i^{\text{ext}}(s)|+|v-v_i(\theta_i^*(s))|+V_i(u,v,\theta_i^*(s))\Big)\Big\} = \\ = \alpha_i(s)H_i(\psi_i^{\text{ext}}(s),\theta_i^*(s)) + \big\langle Q_i(\psi_i^{\text{ext}}(s),\theta_i^*(s)),\beta_i(s)\big\rangle + \big\langle\sigma_i(s),\beta_i(s)\big\rangle - \\ \lambda_i z_i^{-2}a_i(\text{dist}(r_i(\theta_i^*(s)),C))^2 - \alpha_i(s)\lambda_i\varepsilon_i V_i(\theta_i^*(s)) \ \text{ a.a. } s\in[0,1], \end{cases} \tag{6.36}$$

$$\begin{cases} \alpha_i(s)\dfrac{\partial H_i}{\partial u}(\psi_i^{\text{ext}}(s),\theta_i^*(s)) + \dfrac{\partial}{\partial u}\Big\langle Q_i(\psi_i^{\text{ext}}(s),\theta_i^*(s)),\beta_i(s)\Big\rangle \\ \in \dfrac{\partial r_i^*}{\partial u}(\theta_i^*(s))\eta_i^{\text{ext}}(s) + \alpha_i(s)\lambda_i\varepsilon_i\Big(\dfrac{\partial V_i}{\partial u}(\theta_i^*(s)) + B_{\mathbb{R}^m}\Big) \\ \text{a.a. } s:\ |u_i^{\text{ext}}(s)| < c, \end{cases} \tag{6.37}$$

$$|\lambda_i| + \max_{s\in[0,1]}|\psi_i^{\text{ext}}(s)| + \rho_i = 1. \tag{6.38}$$

Here, $\theta_i^* = \pi_i^{*-1}$, $\psi_i^{\text{ext}}(s) = \psi_i^x(\theta_i^*(s))$, $\xi_i(s) = \psi_i^z(\theta_i^*(s))$, $\sigma_i(s) = \psi_i^\zeta(\theta_i^*(s))$, $\zeta_i^{\text{ext}}(s) = \zeta_i(\theta_i^*(s))$, $u_i^{\text{ext}}(s) = u_i(\theta_i^*(s))$,

$$\alpha_i(s) = \frac{a_i}{m_i(\theta_i^*(s))},\quad \beta_i(s) = \frac{a_i v_i(\theta_i^*(s))}{m_i(\theta_i^*(s))},\quad a_i = t_1 - t_0 + \|m_i\|_{\text{L}_1},$$

$\rho_i = 2\lambda_i z_{0,i}^{-2}\text{dist}(p_i,S)$, and

$$\eta_i^{\text{ext}}(s) = \alpha_i(s)\cdot 2\lambda_i z_i^{-2}\text{dist}(r_i(\theta_i^*(s)),C)m_i(\theta_i^*(s))\gamma_i(\theta_i^*(s)),$$

where $\gamma_i(t) \in \text{conv}\partial\text{dist}(r_i(t),C)$ for a.a. t—some measurable selection.

Above and in that which follows: If some function which depends on (x,ζ,u,v) has index i and, at the same time, some of these arguments are omitted, then this means that the values $x_i(\theta_i^*(s))$, $\zeta_i(\theta_i^*(s))$, $u_i(\theta_i^*(s))$, $v_i(\theta_i^*(s))$ replace the omitted arguments, that is, w.r.t. the just performed time variable change. For example,

$$H_i(u,\psi_i^{\text{ext}}(s),\theta_i^*(s)) = H(x_i(\theta_i^*(s)),u,\psi_i^{\text{ext}}(s),\theta_i^*(s)).$$

At the same time, the dependence on the variable ψ is maintained for convenience.

Since $\int_{t_0}^{t_1} V_i(t)dt \to 0$, then it can be proved that, after extracting a subsequence, the following convergence holds true $\theta_i^*(s) \rightrightarrows \hat{\theta}(s)$, $u_i^{\text{ext}}(s) \to \hat{u}^{\text{ext}}(s)$ for a.a. s, and, as a corollary, $x_i^{\text{ext}}(s) := x_i(\theta_i^*(s)) \rightrightarrows \hat{x}^{\text{ext}}(s)$. (See Remark 6.1). Moreover, we may consider that $\alpha_i(s) \xrightarrow{w} \hat{\alpha}(s)$, $\beta_i(s) \xrightarrow{w} \hat{\beta}(s)$ weakly in L_2 due to the uniform boundedness of functions α_i, β_i.

Let us ensure that the sequence of functions $\eta_i^{\text{ext}}(s)$ has a subsequence which is uniformly bounded w.r.t. the L_∞-norm. When sequence $\{\lambda_i z_i^{-2}\}$ is bounded, this fact is obvious due to the construction and the above definitions because functions

$\text{dist}(r_i(\theta_i^*(s)), C)$ are uniformly bounded. Moreover, $\text{dist}(r_i(\theta_i^*(s)), C) \to 0$ for a.a. $s \in [0, 1]$, and then, $\eta_i^{\text{ext}}(s) \to 0$ for a.a. $s \in [0, 1]$. This fact later will be used later.

Therefore, let us assume that either $\lambda_i z_i^{-2} \to \infty$ as $i \to \infty$, or there is a subsequence with such a property, to which we immediately pass. Then, we show that there is a constant $\kappa > 0$ such that

$$|\eta_i^{\text{ext}}(s)| \le \kappa(\lambda_i + |\psi_i^{\text{ext}}(s)|) \ \text{a.a.}\, s \in [0, 1] \ \forall i. \tag{6.39}$$

From the regularity of the optimal trajectory w.r.t. the mixed constraints and from Lemma 6.4, though applied to constraint system $r(x, u, \theta) \in C$ and w.r.t. trajectory $(\hat{x}^{\text{ext}}(s), \hat{\theta}(s))$; see Exercise 6.7 for details, also due to the fact that $x_i^{\text{ext}}(s) \rightrightarrows \hat{x}^{\text{ext}}(s)$, and $\theta_i^*(s) \rightrightarrows \hat{\theta}(s)$, it follows that there exist a number N and measurable functions $\tilde{u}_i(s)$ such that $\|\tilde{u}_i\|_{L_\infty} < c$ and the following inclusion is satisfied:

$$r(x_i^{\text{ext}}(s), \tilde{u}_i(s), \theta_i^*(s)) \in C \ \text{a.a.}\, s \in [0, 1], \ \forall i \ge N.$$

By setting $u = \tilde{u}_i(s)$, $v = v_i(\theta_i^*(s))$ for a.a. s in the left hand of (6.36), and by using (6.38), we obtain

$$\lambda_i z_i^{-2}(\text{dist}(r_i(\theta_i^*(s)), C))^2 \le \text{const}\Big(\lambda_i\varepsilon_i + |\psi_i^{\text{ext}}(s)|\Big) \le \text{const}.$$

Since $\lambda_i z_i^{-2} \to \infty$, it follows that

$$\text{dist}(r_i(\theta_i^*(s)), C) \to 0 \ \text{a.a.}\, s \tag{6.40}$$

uniformly w.r.t. $s \in [0, 1]$. Therefore, in view of the choice of the number c and the continuity of r w.r.t. x, for all sufficiently large i, we have $|u_i^{\text{ext}}(s)| < c$ for a.a. $s \in [0, 1]$.

The next step is to consider $s \in [0, 1]$, and to apply the Lagrange principle (see Theorem 5.5 in [17]) to the nonsmooth problem

$$\begin{aligned}\alpha_i(s)H_i(u, \psi_i^{\text{ext}}(s), \theta_i^*(s)) + \langle Q_i(u, \psi_i^{\text{ext}}(s), \theta_i^*(s)), \beta_i(s)\rangle \\ -\lambda_i z_i^{-2}a_i(\text{dist}(r_i(u, \theta_i^*(s)), C))^2 \\ -\lambda_i\varepsilon_i\alpha_i(s)\Big(|u - u_i^{\text{ext}}(s)| + V_i(u, \theta_i^*(s))\Big) \to \max, \quad |u| \le c.\end{aligned}$$

In view of the maximum condition (6.36), the solution to this problem is precisely the point $u = u_i^{\text{ext}}(s)$. Therefore,

$$\begin{aligned}\alpha_i(s)\frac{\partial H_i}{\partial u}(\psi_i^{\text{ext}}(s), \theta_i^*(s)) + \frac{\partial}{\partial u}\langle Q_i(\psi_i^{\text{ext}}(s), \theta_i^*(s)), \beta_i(s)\rangle \in \\ 2\lambda_i z_i^{-2}a_i\text{dist}(r_i(\theta_i^*(s)), C)\frac{\partial r_i^*}{\partial u}(\theta_i^*(s))\partial\text{dist}(r_i(\theta_i^*(s)), C) + \lambda_i\varepsilon_i a_i(1 + 4c)B_{\mathbb{R}^m}.\end{aligned}$$

Then, by applying the measurable selection lemma,

$$\begin{aligned}\alpha_i(s)\frac{\partial H_i}{\partial u}(\psi_i^{\text{ext}}(s), \theta_i^*(s)) + \frac{\partial}{\partial u}\langle Q_i(\psi_i^{\text{ext}}(s), \theta_i^*(s)), \beta_i(s)\rangle \in \\ \frac{\partial r_i^*}{\partial u}(\theta_i^*(s))\omega_i(s) + \lambda_i\varepsilon_i a_i(1+4c)B_{\mathbb{R}^m},\end{aligned} \tag{6.41}$$

where

$$\omega_i(s) := 2\lambda_i z_i^{-2} a_i \text{dist}(r_i(\theta_i^*(s)), C) n_i(s),$$

being $n_i(s) \in \partial\text{dist}(r_i(\theta_i^*(s)), C)$ some measurable map. From the properties of the subdifferential of the distance function, $|n_i(s)| \leq 1$ for a.a. $s \in [0, 1]$, and, by Theorem 1.105 from [16], $|n_i(s)| = 1$ for a.a. s such that $r_i(\theta_i^*(s)) \notin C$. On the other hand, in view of definition of $\omega_i(t)$, it follows that $\omega_i(s) = \eta_i^{\text{ext}}(s) = 0$, a.a. s such that $r_i(\theta_i^*(s)) \in C$. Moreover, $|\gamma_i(\theta_i^*(s))| \leq 1$, a.a. $s \in [0, 1]$, whence $|\eta_i^{\text{ext}}(s)| \leq |\omega_i(s)|$ for a.a. $s \in [0, 1]$.

By Theorem 1.105 from [16], we have

$$n_i(s) \in \bigcup_{y \in \Pi_C(r_i(\theta_i^*(s)))} N_C(y) \text{ a.a. } s \in [0, 1] \text{ such that } r_i(\theta_i^*(s)) \notin C. \tag{6.42}$$

By taking into account the definition of ω_i, and also (6.40), and by using that $|\eta_i^{\text{ext}}(s)| \leq |\omega_i(s)|$ for a.a. $s \in [0, 1]$, the existence of number κ and Estimate (6.39) follow directly from Conditions (6.24), (6.41), and (6.42).

From (6.38) and (6.39), it follows that the functions $\eta_i^{\text{ext}}(s)$ are uniformly bounded up to a subsequence which will be considered from this point.

Let us show that the auxiliary conjugate functions ξ_i, σ_i converge uniformly to zero. Concerning σ_i this fact is obvious since $\zeta_i - \bar{\zeta}_i \rightrightarrows 0$ due to $\|V_i\|_{L_1} \to 0$ and Condition (6.38). Regarding ξ_i, it is enough to show that $\xi_i(0) \to 0$.

For this, in view of (6.34), it is sufficient to establish that

$$\int_0^1 2\lambda_i z_i^{-3} \text{dist}(r_i(\theta_i^*(s)), C))^2 ds \to 0.$$

From the definition of $\eta_i^{\text{ext}}(s)$, we have

$$2\lambda_i \int_0^1 z_i^{-3} \text{dist}(r_i(\theta_i^*(s)), C))^2 ds = a_i^{-1} \int_0^1 z_i^{-1} \text{dist}(r_i(\theta_i^*(s)), C)) |\eta_i^{\text{ext}}(s)| ds.$$

Therefore, since $\|\eta_i^{\text{ext}}\|_{L_\infty} \leq \text{const}$, it is enough to ensure that $\text{dist}(r_i(\theta_i^*(s)), C) z_i^{-1} \to 0$ in L_1. However, this sequence converges even in L_2 due to (6.31) and the variable change. So, $\xi_i(0) \to 0$.

From (6.34), (6.38), and in view of $\|\eta_i^{\text{ext}}\|_{L_\infty} \leq \text{const}$, it follows that the functions $\psi_i^{\text{ext}}(s)$ are equicontinuous and uniformly bounded. Thus, by means of the Arzela–

Ascoli theorem and in view of weak sequential compactness of the unit ball in $L_2([0, 1]; \mathbb{R}^d)$, we have, after extracting a subsequence, that there exist λ, ψ^{ext}, η^{ext} such that $\lambda_i \to \lambda$, $\psi_i^{ext} \rightrightarrows \psi^{ext}$, and $\eta_i^{ext} \xrightarrow{w} \eta^{ext}$ weakly in L_2.

By definition of η_i^{ext} and by using their weak convergence and (6.39), it follows that, as $i \to \infty$,

$$\eta^{ext}(s) \in \text{conv} N_C(r^{ext}(s)),$$

$$|\eta^{ext}(s)| \leq \kappa(\lambda + |\psi^{ext}(s)|) \text{ a.a. } s \in [0, 1]. \tag{6.43}$$

(Note that if sequence $\{\lambda_i z_i^{-2}\}$ is bounded, then (6.43) is true for any $\kappa > 0$ as in this case $\eta^{ext} = 0$.)

By passing to the limit in (6.34) as $i \to \infty$, we obtain Condition (6.26), where Exercise (4.4) of Chap. 2 has also been used.

The passage to the limit in the transversality condition (6.35) is straightforward. For this, we use the fact that $\xi_i(0) \to 0$, the fact that, in view of (6.31), $z_i^{-1}\text{dist}(p_i, S) \to 0$, the results on subdifferentiation of the distance function collected in [16] (Sect. 1.3.3, see Theorems 1.97 and 1.105 therein), the upper semicontinuity of the cone $N_S(p)$, and also (6.38). Then, as $i \to \infty$, we derive (6.27).

Let us prove (6.28) and (6.29). By virtue of (6.31) and of the change of time variable, we have that

$$\int_0^1 z_i^{-2}\text{dist}(r_i(\theta_i^*(s)), C))^2 ds \to 0,$$

and, by virtue of (6.32),

$$\int_0^1 \frac{|v_i(\theta_i^*(s)) - \bar{v}_i(\theta_i^*(s))|}{m_i(\theta_i^*(s))} ds = a_i^{-1} \int_{t_0}^{t_1} |v_i(t) - \bar{v}_i(t)| dt \to 0.$$

By using this, according to the chosen number κ_i, we conclude that

$$\ell(A) = 1, \text{ where } A = \left\{ s \in [0, 1] : \liminf_{i \to \infty} \kappa_i^{-1} |v_i(\theta_i^*(s))| = 0 \right\}.$$

Therefore, for a.a. s, there exists a subsequence such that

$$z_i^{-2}\text{dist}(r_i(\theta_i^*(s)), C))^2 \to 0 \text{ and } \kappa_i^{-1}|v_i(\theta_i^*(s))| \to 0.$$

Note that from the last condition, it follows that $\kappa_i \alpha_i(s) \to \infty$. Take $s \in [0, 1]$ with the above-mentioned property such that $u_i^{ext}(s) \to \hat{u}^{ext}(s)$ (the set of such s has full measure). Take an arbitrary vector $v \in K \cap B_{\mathbb{R}^k}$. Substitute vector $\kappa_i v$ in the left hand of (6.36) and divide both parts of (6.36) by $\kappa_i \alpha_i(s)$. Then, using the already established facts and by extracting the appropriate subsequence, as $i \to \infty$, we obtain

$$\max_{u \in U^{\text{ext}}(s)} \left\langle Q(\hat{x}^{\text{ext}}(s), u, \psi^{\text{ext}}(s), \hat{\theta}(s)), v \right\rangle \leq 0. \tag{6.44}$$

Here, we also used that $\sigma_i(s) \rightrightarrows 0$ and the fact that according to the continuity properties of mapping $U(x, \theta)$ w.r.t. x, θ, for every $u \in U^{\text{ext}}(s)$, there exists a sequence of vectors $\tilde{u}_i \in \mathbb{R}^m$ satisfying $r_i(\tilde{u}_i, \theta_i^*(s)) \in C$, such that $\tilde{u}_i \to u$.

Substituting in the left hand of (6.36) values $u = u_i^{\text{ext}}(s)$ and $v = 0$, integrating and passing to the limit as $i \to \infty$, we obtain

$$\int_0^1 \left\langle \hat{Q}^{\text{ext}}(s), \hat{\beta}(s) \right\rangle ds \geq 0.$$

Therefore, from Condition (6.44), which is satisfied for every $v \in K$, we extract that

$$\left\langle \hat{Q}^{\text{ext}}(s), \hat{\beta}(s) \right\rangle = 0 \ \text{ a.a. } s \in [0, 1].$$

Thus, Condition (6.29) is proved.

In order to prove (6.28), we set $v = 0$ in (6.36). Take any function $u(\cdot)$ such that $u(s) \in U^{\text{ext}}(s)$ a.a. s. Following Lemma 6.4, there exists, for every i, a function $\tilde{u}_i(s)$ with values in $U(x_i^{\text{ext}}(s), \theta_i^*(s))$ such that $\|\tilde{u}_i - u\|_{L_\infty} \to 0$. Let us take $u = \tilde{u}_i(s)$ in the left hand of (6.36) and, then, integrate both parts of it. After passing to the limit as $i \to \infty$, we have

$$\int_0^1 \hat{\alpha}(s) H(\hat{x}^{\text{ext}}(s), u(s), \psi^{\text{ext}}(s), \hat{\theta}(s)) ds \leq \int_0^1 \hat{\alpha}(s) H^{\text{ext}}(s) ds.$$

However, this is precisely Condition (6.28) albeit derived in the equivalent integral form. Thus, (6.28) is proved.

Condition (6.30) can be obtained by integration of (6.37) on the intervals $[0, s]$ and by subsequently passing to the limit under the integral.

It remains to demonstrate the validity of the nontriviality condition (6.25). Note that if $\lambda_i \to 0$ and $\psi_i^{\text{ext}} \rightrightarrows 0$, then inevitably $\rho_i \to 0$. Indeed, this is due to (6.35), the fact that $\xi_i(0) \to 0$, as well as $z_i^{-1}\text{dist}(p_i, S) \to 0$, the definition of ρ_i and also Theorem 1.105 from [16] by virtue of which we have that $|h| = 1 \ \forall h \in \partial N_S(p)$ when $p \notin S$. Therefore, from (6.38), as $i \to \infty$,

$$\lambda + \|\psi^{\text{ext}}\|_C > 0. \tag{6.45}$$

The nontriviality condition (6.25) now follows from (6.26), (6.43), and (6.45) in view of Gronwall's inequality (see, e.g., similar arguments in [6], Chap. 2).

The proof is complete.

Remark 6.2 The condition that the sets $U(t)$, $U_\tau(s)$ are uniformly bounded in the formulation of Theorem 6.1 can be dispensed with for some classes of C, e.g., when C is convex or semialgebraic, through the technique proposed in [4].

Remark 6.3 The regularity of the extended trajectory w.r.t. the mixed constraints can be considerably weakened and, from the global type, can be dropped down to a local type condition imposed in some δ-tube about the minimizer. Then, the maximum condition changes as the maximum should be taken over the closure of the set of regular points. This is proved only for some classes of C, as in Remark 6.2. For details, see [4].

6.6 Exercises

Exercise 6.1 Ensure that for the constraint system

$$u_1^2 + u_2^2 - 1 \geq 0, \quad (u_1 - 2)^2 + u_2^2 - 1 \geq 0$$

the regularity condition from Definition 6.3 is not satisfied. (Here, $C = \mathbb{R}^2_+$). By resolving this constraint system w.r.t. u rewrite it in an equivalent, though geometrical (that is, in terms of sets, not inequalities) form, so that it becomes regular. Use the discussion from Sect. 6.3. Ensure that the same can be carried out w.r.t. to any constraint system where r depends only on u.

Exercise 6.2 Consider the following mixed constraint

$$x_1 u_1^2 - x_2 u_2^2 = 0.$$

Verify that the regularity imposed in Definition 6.2 is violated at the origin $u = (0, 0)$. Ensure that this constraint system cannot be explicitly geometrically resolved w.r.t. u in the framework of the problem formulation in (6.1).

Exercise 6.3 Verify that map $\rho(\cdot, \cdot)$ introduced in this chapter is a metric. Show that metric space $(\mathscr{P}, \rho)$ is not complete. Ensure that it will be complete as soon as conventional control $u(\cdot)$ is L_1-function w.r.t. $\ell + |\vartheta|$.

Exercise 6.4 Verify the validity of Properties (A), (B). Use the discontinuous time variable change and the material of Chap. 4.

Exercise 6.5 Given a sequence of measurable integrable functions f_i such that $\|f_i\|_{L_1} \to 0$, ensure that $\|f_i\|_{L_p} \to 0$, for any $p : 1 < p < \infty$, wherever f_i's are uniformly bounded in the L_∞-norm.

Exercise 6.6 Ensure that function $\frac{1}{\hat{m}_1(t)}$ is integrable, where $\hat{m}_1(t)$ is defined in the proof of Theorem 6.1.

Exercise 6.7 Verify that, under the assumptions imposed on map r, the following simple corollary of Lemma 6.4 (which is, in fact, also an extension of it w.r.t. the time variable) is valid. Let $x_i \in C(T; \mathbb{R}^n)$, $\theta_i \in C(T; \mathbb{R})$, $x_i \rightrightarrows x$, $\theta_i \rightrightarrows \theta$, and $u(t) \in U(x(t), \theta(t))$ for a.a. t, where trajectory $(x(\cdot), \theta(\cdot))$ is regular w.r.t. the mixed constraints $r(x, u, \theta) \in C$, while sets $U(x(t), \theta(t))$ are uniformly bounded w.r.t. $t \in T$. Then, for each sufficiently large i, there exists a measurable function $u_i(\cdot)$ such that $u_i(t) \in U(x_i(t), \theta_i(t))$ for a.a. t, and

$$\|u_i - u\|_{L_\infty} \to 0.$$

Exercise 6.8* Prove Lemma 6.2 when X is a Banach space. Use the Ekeland variational principle and apply the same arguments as in the finite-dimensional case. Ensure that the assertion of lemma is not valid when Y is a Banach space.

Exercise 6.9* Ensure the existence of solution to the extension of the minimal fuel consumption problem proposed in Sect. 6.2.

Exercise 6.10* Embed the extension (2.1) of Chap. 2 within the framework of the extension given by (6.1) of this chapter. Indicate the control policy, the topology, and extra hypotheses under which the extension of Chap. 2 is well posed in the sense given in Introduction. Use metric ρ defined in formula (6.4).

References

1. Arutyunov, A.: Perturbations of extremal problems with constraints and necessary optimality conditions. J. Sov. Math. **54**(6), 1342–1400 (1991)
2. Arutyunov, A., Karamzin, D.: Necessary conditions for a weak minimum in an optimal control problem with mixed constraints. Differ. Equ. **41**(11), 1532–1543 (2005)
3. Arutyunov, A., Karamzin, D., Pereira, F.: Maximum principle in problems with mixed constraints under weak assumptions of regularity. Optimization **59**(7), 1067–1083 (2010)
4. Arutyunov, A., Karamzin, D., Pereira, F., Silva, G.: Investigation of regularity conditions in optimal control problems with geometric mixed constraints. Optimization **65**(1), 185–206 (2016)
5. Arutyunov, A., Zhukovskiy, S.: Local solvability of control systems with mixed constraints. Differ. Equ. **46**(11), 1561–1570 (2010)
6. Arutyunov, A.V.: Optimality conditions. Abnormal and degenerate problems. Mathematics and its Applications. Kluwer Academic Publishers, Dordrecht (2000)
7. Clarke, F., De Pinho, M.: Optimal control problems with mixed constraints. SIAM J. Control Optim. **48**(7), 4500–4524 (2010)
8. Devdariani, E., Ledyaev, Y.: Maximum principle for implicit control systems. Appl. Math. Optim. **40**(1), 79–103 (1999)
9. Dubovitskii, A., Milyutin, A.: Necessary conditions for a weak extremum in optimal control problems with mixed constraints of the inequality type. USSR Comput. Math. Math. Phys. **8**(4), 24–98 (1968)
10. Ekeland, I.: On the variational principle. J. Math. Anal. Appl. **47**, 324–353 (1974)
11. Filippov, A.: On certain problems of optimal regulation. Bulletin of Moscow State University. Ser. Math. Mech. (2), 25–38 (1959)

12. Hestenes, M.: Calculus of Variations and Optimal Control Theory. Wiley, New York (1966)
13. Makowski, K., Neustadt, L.: Optimal control problems with mixed control-phase variable equality and inequality constraints. SIAM J. Control **12**(2), 184–228 (1974)
14. Milyutin, A.: Maximum Principle for General Optimal Control Problem. Fizmatlit, Moscow (2001). [in Russian]
15. Mordukhovich, B.: Maximum principle in the problem of time optimal response with nonsmooth constraints. J. Appl. Math. Mech. **40**, 960–969 (1976)
16. Mordukhovich, B.: Variational analysis and generalized differentiation I. Basic Theory, Grundlehren der Mathematischen Wissenschaften [Fundamental Principles of Mathematical Sciences], vol. 330. Springer, Berlin (2006)
17. Mordukhovich, B.: Variational analysis and generalized differentiation II. Applications, Grundlehren der Mathematischen Wissenschaften [Fundamental Principles of Mathematical Sciences], vol. 331. Springer, Berlin (2006)
18. Neustadt, L.: Optimization. Princeton University Press, Princeton (1976)
19. Páles, Z., Zeidan, V.: Strong Local Optimality Conditions for Control Problems with Mixed State-Control Constraints, pp. 4738–4743 (2002)
20. de Pinho, M., Loewen, P., Silva, G.: A weak maximum principle for optimal control problems with nonsmooth mixed constraints. Set-Valued Var. Anal. **17**(2), 203–221 (2009)
21. de Pinho, M., Vinter, R.: Necessary conditions for optimal control problems involving nonlinear differential algebraic equations. J. Math. Anal. Appl. **212**(2), 493–516 (1997)
22. de Pinho, M., Vinter, R., Zheng, H.: A maximum principle for optimal control problems with mixed constraints. IMA J. Math. Control Inf. **18**(2), 189–205 (2001)
23. Pontryagin, L., Boltyanskii, V., Gamkrelidze, R., Mishchenko, E.: Mathematical theory of optimal processes. In: Neustadt, L.W. (ed.) Translate from the Russian, 1st edn. Interscience Publishers, Wiley (1962)
24. Robinson, S.M.: Regularity and stability for convex multivalued functions. Math. Oper. Res. **1**(2), 130–143 (1976)

Chapter 7
General Nonlinear Impulsive Control Problems

Abstract In this concluding chapter, an extension of the classical control problem is given in the most general nonlinear case. The essential matter is that now the control variable is not split into conventional and impulsive types, while the dependence on this unified control variable is not necessarily affine. By combining the two approaches, the one based on the Lebesgue discontinuous time variable change, and the other based on the convexification of the problem by virtue of the generalized controls proposed by Gamkrelidze, a fairly general extension of the optimal control problem is constructed founded on the concept of generalized impulsive control. A generalized Filippov-like existence theorem for a solution is proved. The Pontryagin maximum principle for the generalized impulsive control problem with state constraints is presented. Within the framework of the proposed approach, a number of classic examples of essentially nonlinear problems of calculus of variations which allow for discontinuous optimal arcs are also examined. The chapter ends with seven exercises.

7.1 Introduction

Finally, we come to the question of a well-posed extension for a general optimal control problem, not necessarily linear w.r.t. v-variable, that is, the case of the dynamics defined by (8) in the Introduction. Despite the generality of this question, such extension is relevant in view of a variety of classical examples, which are essentially nonlinear and where discontinuous solutions arise representing some physical meaning. It follows then that discontinuous solutions to nonlinear problems w.r.t. v are still interesting from the point of view of applications. Let us review a few such examples arising in calculus of variations.

Consider the following well-known problem (proposed by L. Euler):

© Springer Nature Switzerland AG 2019

A. Arutyunov et al., *Optimal Impulsive Control*, Lecture Notes in Control and Information Sciences 477, https://doi.org/10.1007/978-3-030-02260-0_7

$$\begin{aligned} &\text{Minimize} \int_0^1 x(t)\sqrt{1+(\dot{x}(t))^2}dt, \\ &\text{subject to: } x(0)=r_1,\ x(1)=r_2, \\ &\qquad\qquad x(t)\ge 0. \end{aligned} \tag{7.1}$$

This is the so-called minimal surface problem. Physically, the solution $x(t)$ is the shape of a soap bubble or a membrane stretched over two parallel disks with radiuses r_1 and r_2. The application of the Euler–Lagrange principle leads to a second-order differential equation and to a boundary problem, which does not have solutions for certain values of r_1, r_2. The physical meaning is as follows: If numbers r_1, r_2 are sufficiently large relative to the distance between the disks, the membrane exists and the surface of revolution is smooth. However, if we increase the distance between disks, the soap bubble stretches and, at some point, bursts: At that precise moment, the smooth and continuous solution fails to exist. Nonetheless, this does not mean that a solution does not exist at all. In this degenerate case, the solution is $x(0) = r_1$, $x(1) = r_2$, $x(t) = 0$, $t \in (0, 1)$, and thus, it exhibits discontinuities.

Consider another example = the so-called Dido problem:

$$\begin{aligned} &\text{Minimize} \int_{-1}^1 x(t)dt, \\ &\text{subject to: } x(-1)=x(1)=0,\ \ x(t)\ge 0, \\ &\qquad\qquad \int_{-1}^1 \sqrt{1+(\dot{x}(t))^2}dt = L. \end{aligned} \tag{7.2}$$

Once again, a continuous solution fails to exist when the length of the arc L is sufficiently great. The Dido problem is a typical example of the so-called isoperimetric problem. The situation in which there is no solution is fairly common w.r.t. such kind of problems.

The isoperimetric version of the Euler problem is the catenary[1]:

$$\begin{aligned} &\text{Minimize} \int_0^1 x(t)\sqrt{1+(\dot{x}(t))^2}dt, \\ &\text{subject to: } x(0)=r_1,\ x(1)=r_2,\ x(t)\ge 0, \\ &\qquad\qquad \int_0^1 \sqrt{1+(\dot{x}(t))^2}dt = L. \end{aligned} \tag{7.3}$$

[1] A small historical note: The equation of the catenary curve was derived by Leibniz, Huygens, and Johann Bernoulli in 1691. They were the first to discover that this curve is a hyperbolic cosine, but not a parabola, as had been thought before. The shape of the soap bubble in (7.1) is, as we see from (7.3), also formed by the catenary. (This was noticed by Euler.)

This problem again naturally allows discontinuous solutions.

Another example of the isoperimetric problem consists in minimizing the norm of a function in L_1 over the elements of the unit sphere in L_2.

$$\begin{aligned} &\text{Minimize} \int_0^1 |\dot{x}(t)|dt, \\ &\text{subject to} \;\; x(0) = 0, \;\; \int_0^1 |\dot{x}(t)|^2 dt = 1. \end{aligned} \tag{7.4}$$

A continuous solution does not exist here. Indeed, the infimum is zero, but is not reached due to the integral constraint. There are many other simple examples where such a property manifests and solutions become discontinuous.

In the realm of this context, the aim of this chapter is to provide strict mathematical meaning to discontinuous solutions which may arise in the following optimal control problem with constraints:

$$\begin{aligned} \text{Minimize} \;\; & \varphi(p), \\ \text{subject to:} \;\; & \dot{x} = g(x, v, t), \quad t \in [t_0, t_1], \\ & p = (x(t_0), x(t_1)) \in S, \\ & v(t) \in V \text{ a.a. } t \in [t_0, t_1], \\ & l(x(t), t) \leq 0 \;\; \forall t \in [t_0, t_1], \end{aligned} \tag{7.5}$$

where

$$g : \mathbb{R}^n \times \mathbb{R}^k \times \mathbb{R}^1 \to \mathbb{R}^n, \quad l : \mathbb{R}^n \times \mathbb{R}^1 \to \mathbb{R}^d$$

are given continuous maps, S is a given closed subset of $\mathbb{R}^{2n}$, V is a given closed subset of $\mathbb{R}^k$, *not necessarily bounded*, and $v(t)$ is a control function. The function l defines the state constraints which were studied in Chap. 5. This function appears in the problem formulation due to the fact that the Euler problem, the Dido problem considered above as well as many others, essentially involves state constraints.

Note that Problem (7.5) is more general in its statement than Problem (3), in which the feasible control set U is compact. Indeed, the set V in (7.5) is permitted to be unbounded. The problem formulation in (7.5) also provides the opportunity to minimize integrals

$$\int_{t_0}^{t_1} g_0(x, v, t)dt,$$

and, therefore, includes the above-considered examples. This can be implemented by virtue of the extra state variable χ with $\dot{\chi} = g_0(x, v, t)$, $\chi_0 = 0$, and by subsequently minimizing χ_1. Also note that, by setting $v = \dot{x}$, the above examples (7.1)–(7.4) fall into the formulation (7.5) with $V = \mathbb{R}$.

In order to achieve the stated objectives, the approach to extension proposed by Gamkrelidze in [4, 5] is exploited. This approach is based on the notion of generalized control; see [6, 8]. The extension in this chapter is a certain upgrade of that extension, though related to *discontinuous arcs*, unlike the case of continuous arcs considered in [6]. The underlying idea consists in bonding the conventional generalized control, and the Borel measure given on the time interval, through the Lebesgue discontinuous variable change. The result of such bonding is the generalized impulsive control. The constructed extension generalizes the extensions given in the previous chapters, which were designed for linear dynamics w.r.t. v-variable with separated control variables u, v.

7.2 Preliminaries

To construct the extension, we will need a natural compactification of space $\mathbb{R}^k$, which is the result of union of $\mathbb{R}^k$ and the set S_∞ called "infinity sphere." Infinity sphere is a conventional $(k-1)$-dimensional unit sphere, but it lies in the counterpart space $\mathbb{R}^k$. Formally, such a compactification is defined as pair (Θ, B_1), where B_1 is the closed unit ball in the $\mathbb{R}^k$-counterpart and $\Theta : \mathbb{R}^k \to B_1$ is an embedding which is defined by the formula:

$$\Theta(v) = \frac{v}{1+|v|}, \quad v \in \mathbb{R}^k.$$

By extending this embedding onto the infinite sphere by the identity mapping $\Theta(e) = e$, $e \in S_\infty$, we obtain a topology on $\bar{\mathbb{R}}^k := \mathbb{R}^k \cup S_\infty$, in which the open sets are those sets $\Theta^{-1}(O)$, where O is open in the induced topology on B_1 (see Exercise 7.1). Then, $\Theta : \bar{\mathbb{R}}^k \to B_1$ is a homeomorphism, and space $\bar{\mathbb{R}}^k$ is topologically equivalent to the closed unit ball and, therefore, is compact.

The feasible set V is closed in $\mathbb{R}^k$, but not necessarily in $\bar{\mathbb{R}}^k$. Denote by $\bar{V}$ its closure in the just described topology of compact $\bar{\mathbb{R}}^k$. Clearly, we have the relation

$$\bar{V} = \Theta^{-1}\Big(\mathrm{cl}\,\Theta(V)\Big),$$

which will be implicitly used in what follows.

Let us associate the control problem (7.5) with some a priori given scalar function $\omega : [0, +\infty] \to [0, +\infty]$ which satisfies the following properties.

(1) $\omega(0) = 0$.
(2) continuous and increasing.
(3) $\lim_{\rho\to\infty} \omega(\rho) = +\infty$.[2]

[2]It is also possible to consider the case of finite limit. But then, as it will become clear from the forthcoming exposition, the trajectory discontinuities will not arise in the extension. Therefore, this case is not of interest for this book. (It is already encompassed in the known theory.)

Function ω plays a crucial role in the extension of (7.5). Its purpose is to characterize the growth of the dynamics at infinity. We will consider that the measurable control function $v(t)$ in Problem (7.5) is such that function $\omega(|v(t)|)$ is integrable. So, if, e.g., $\omega(\rho) = \rho$, then v is a function in L_1, if $\omega(\rho) = \rho^2$, then v is in L_2, and so on. Thereby, function ω defines the base class of measurable feasible controls in Problem (7.5). Those functions v, for which $\omega(|v|)$ is nonintegrable, are not involved in the extension construction. If it is necessary to extend a problem with bounded controls, i.e., of class L_∞, then ω can be selected arbitrarily by taking into account the above-suggested properties (1), (2), (3), for example, $\omega(\rho) = \rho$.[3]

Consider mapping

$$\bar{g}(x, v, t) := \frac{g(x, v, t)}{1 + \omega(|v|)}.$$

Our basic assumption in what follows is that the mapping $\bar{g}$ is *continuously extendable* from $\mathbb{R}^n \times \mathbb{R}^k \times \mathbb{R}^1$ onto $\mathbb{R}^n \times \bar{\mathbb{R}}^k \times \mathbb{R}^1$. This means that for every x, t mapping $\bar{g}(x, \cdot, t)$ is defined over the compactified space $\bar{\mathbb{R}}^k$ and is continuous. In particular, it is continuous on the infinity sphere, and

$$\bar{g}(x, e, t) = \lim_{v \to e} \frac{g(x, v, t)}{1 + \omega(|v|)} \quad \forall x, t \in \mathbb{R}^n \times \mathbb{R}^1, \quad \forall e \in S_\infty.$$

Here, the convergence of points $v \in \mathbb{R}^k$ to point $e \in S_\infty$ is understood in the sense of topology on $\bar{\mathbb{R}}^k$ which implies that

$$\Theta(e) = \frac{v}{|v|} + \alpha(v),$$

where $\alpha(v) \to 0$ as $|v| \to \infty$, $v \neq 0$.

We shall use the following definition.

Definition 7.1 The control problem (7.5) is said to allow the impulsive extension of order ω, provided that the map $\bar{g}$ is continuously extendable onto $\mathbb{R}^n \times \bar{\mathbb{R}}^k \times \mathbb{R}^1$ and is nontrivial on the infinite sphere, that is

$$\bar{g}(\cdot, \cdot, \cdot)|_{\mathbb{R}^n \times S_\infty \times \mathbb{R}^1} \neq 0.$$

Thus, if there are points x, t such that the map $\bar{g}(x, \cdot, t) : S_\infty \to \mathbb{R}^n$ is not equivalent to zero map, then the control problem allows the impulsive extension of order ω. It is simple to see that the Euler and Dido problems (7.1)–(7.3) allow the extensions of linear order ρ while Problem (7.4) already requires quadratic growth of ω: $\omega(\rho) = \rho^2$.

We shall use the following assumptions w.r.t. $\bar{g}$.

(H1) Mapping $\bar{g}$ satisfies the following estimate:

---

[3]Note that the solution to the extended problem in this case may happen to lie in L_1. Then, the transition to discontinuous trajectories is already redundant.

$$|\bar{g}(x, v, t)| \leq \kappa(t)(1 + |x|) \ \forall (x, v, t) \in \mathbb{R}^n \times \bar{\mathbb{R}}^k \times \mathbb{R}^1,$$

where κ is some integrable on $[t_0, t_1]$ function.

(H2) Mapping $\bar{g}$ is continuously differentiable w.r.t. x, t for all $v \in \bar{\mathbb{R}}^k$, while its partial derivatives w.r.t. x, t are continuous on $\mathbb{R}^n \times \bar{\mathbb{R}}^k \times \mathbb{R}^1$.

Consider a scalar Borel measure $\mu : \mathscr{B}(T) \to [0, +\infty)$, $T = [t_0, t_1]$. Here, $\mathscr{B}(T)$ stands for the σ-algebra of Borel subsets of T.

Denote by $\mathscr{D}(t; \mu)$ the following Radon–Nikodym derivative:

$$\mathscr{D}(t; \mu) := \frac{d\ell}{d(\ell + \mu)}, \tag{7.6}$$

where ℓ is, as usual, the Lebesgue measure on the real line (length). Note that $\mathscr{D}(\cdot; \mu)$ is $(\ell + \mu)$-measurable and takes values in $[0, 1]$.

As in the previous chapters, consider the discontinuous time variable change $\pi : T \to [0, c]$,

$$\pi(t) = t - t_0 + \mu([t_0, t]), \ t \in (t_0, t_1], \ \ \pi(t_0) = 0,$$

where $c = t_1 - t_0 + \|\mu\|$, and $\|\mu\| = \mu([t_0, t_1])$ is the total variation.

There exists the inverse function $\theta(s)$: $[0, c] \to T$ such that

(a) $\theta(s)$ is increasing.

(b) $\theta(s)$ is absolutely continuous and, moreover, satisfies the Lipschitz condition: $|\theta(s) - \theta(t)| \leq |s - t| \ \forall s, t$.

(c) $\theta(s) = \tau, \forall s \in \Gamma_\tau, \forall \tau \in T$, where $\Gamma_\tau = [\pi(\tau^-), \pi(\tau^+)]$.

Recall that function $v(\theta(s))$ is measurable provided that function $v(t)$ is $(\ell + \mu)$-measurable. Then, the following variable change in the integral makes sense:

$$\int_{t_0}^{t_1} v(t) d\mu = \int_0^c v(\theta(s)) m(\theta(s)) ds,$$

where $m(t)$ stands for the Radon–Nikodym derivative of measure μ w.r.t. $\ell + \mu$.

Finally, we recall that due to Gamkrelidze [6], the generalized control is a weakly measurable family of probabilistic Radon measures $\nu_t : \mathscr{B}(V) \to [0, 1]$, $t \in [t_0, t_1]$. "Weakly measurable" means that for any continuous scalar function $h(v, t)$ the function

$$\left\langle h(v, t), \nu_t \right\rangle := \int_V h(v, t) d\nu_t$$

is measurable w.r.t. t. Similarly, "weakly μ-measurable" means that the above function is μ-measurable. If the set V is compact, then the set of generalized controls is weakly sequentially compact [6]. However, when the set V is unbounded this is not so. The generalized controls and their properties are broadly used in the forthcoming analysis.

7.3 Extension Concept

By using the new objects introduced in Sect. 7.2, let us describe the extended problem. The extension for Problem (7.5) acquires the form:

$$\text{Minimize } \varphi(p), \tag{7.7}$$

$$\text{subject to: } dx = \langle \bar{g}(x, v, t), d\wp \rangle, \tag{7.8}$$

$$p = (x_0, x_1) \in S, \tag{7.9}$$

$$l(x, t) \le 0, \tag{7.10}$$

$$\wp = \{\mu, \nu_t, \nu_s^\tau\}, \quad \text{supp}(\wp) \subseteq \bar{V}.$$

The above formulas and notations require clarification. The necessary definitions are outlined below. The symbol $\wp$ designates the *generalized impulsive control*. By definition, it is constituted by the following three components:

- $\mu : \mathcal{B}(T) \to [0, +\infty)$—a nonnegative scalar Borel measure;
- $\nu_t : \mathcal{B}(\bar{V}) \times T \to [0, 1]$—a weakly $(\ell + \mu)$-measurable family of Radon probabilistic measures defined on $\bar{V}$, which depends on $t \in [t_0, t_1]$, such that

$$\mathcal{D}(t; \mu) + \int_{\bar{V}} \Omega(v) d\nu_t = 1 \text{ a.a. } t \text{ w.r.t. } \ell + \mu, \tag{7.11}$$

where

$$\Omega(v) = \begin{cases} \dfrac{\omega(|v|)}{1 + \omega(|v|)}, & \text{when } v \in \mathbb{R}^k, \\ 1, & \text{when } v \in S_\infty; \end{cases}$$

- $\nu_s^\tau : \mathcal{B}(\bar{V} \cap S_\infty) \times [0, 1] \times \text{Ds}(\mu) \to [0, 1]$—a family of Radon probabilistic measures defined on the infinite sphere, which depends on $s \in [0, 1]$ and also on $\tau \in \text{Ds}(\mu)$, weakly ℓ-measurable w.r.t. s for each $\tau \in \text{Ds}(\mu)$.

Above, Borel measure μ is identified with its unique Lebesgue–Stieltjes extension. Therefore, "μ-measurable" means measurability in the sense of Lebesgue.

As can be seen, the family of measures $\nu_t := \{\nu_t\}_{t \in T}$ and the family of measures $\nu_s^\tau := \{\nu_s^\tau\}_{s \in [0,1]}$, for any fixed $\tau \in \text{Ds}(\mu)$, are generalized controls. The family of generalized controls $\{\nu_s^\tau\}_{\tau \in \text{Ds}(\mu)}$ is called *attached* to the control measure μ. Note that generalized control ν_t is also connected to the control measure μ. Indeed, besides the μ-measurability, the validity of Condition (7.11) is also required.

The symbolic notation $\text{supp}(\wp) \subseteq \bar{V}$ refers to the above conditions imposed on the supports of ν_t, ν_s^τ.

Let us now proceed to the concept of trajectory. Denote by $x_\tau(s)$ the solution to the attached differential system:

$$\begin{cases} \dot{x}_\tau(s) = \Delta_\tau \cdot \langle \bar{g}(x_\tau(s), v, \tau), \nu_s^\tau \rangle, \ s \in [0, 1], \\ x_\tau(0) = x(\tau^-), \end{cases}$$

where $\Delta_\tau := \mu(\{\tau\})$.

The function of bounded variation $x(t)$ is said to be solution to the differential equation (7.8) corresponding to the initial value x_0, provided that:

$$x(t) = x_0 + \int_{t_0}^{t} \Big\langle \bar{g}(x, v, \varsigma), \nu_\varsigma \Big\rangle d(\ell + \mu_c) + \sum_{\tau \in \text{Ds}(\mu):\, \tau \le t} \Big(x_\tau(1) - x_\tau(0) \Big)$$

for all $t \in (t_0, t_1]$ and $x(t_0) = x_0$. Above, μ_c, as usual, stands for the continuous component of the measure.

It remains to define the meaning of the state constraints (7.10). Note that inequality $l(x, t) \le 0$ should be understood in a broader sense than that of conventional inequality. This is due to the presence of the attached family of trajectories $x_\tau(s)$. Namely, the extended trajectory $(x(t), \{x_\tau(s)\}_{\tau \in \text{Ds}(\mu)})$ satisfies the state constraints iff:

$$l(x, t) \le 0 \quad \Leftrightarrow \quad \begin{cases} l(x(t), t) \le 0, \quad \forall t \in T, \\ l(x_\tau(s), \tau) \le 0, \ \forall s \in [0, 1], \ \forall \tau \in \text{Ds}(\mu). \end{cases}$$

The pair $(x, \wp)$ is called control process if (7.8) is satisfied. A control process is said to be feasible provided that endpoint constraints (7.9) and state constraints (7.10) are satisfied. Let us denote the set of all feasible processes by $\mathscr{C}$. A feasible process $(\hat{x}, \hat{\wp}) \in \mathscr{C}$ is said to be optimal, or solution, if the value of the integral in (7.7) on the element $(\hat{x}, \hat{\wp})$ is the least possible finite value over set $\mathscr{C}$.

Let us comment on the given definitions. Problem (7.7) represents a true extension of Problem (7.5) as for any feasible control $v(t)$ of Problem (7.5) there exists a control $\wp$ in Problem (7.7) such that the corresponding trajectories and, consequently, the values of the cost functional coincide. Indeed, let $v(t)$ be a control function. Consider the absolutely continuous Borel measure

$$\mu(C) = \int_C \omega(|v(t)|)dt, \quad C \in \mathscr{B}(T),$$

and set $\nu_t = \delta_{v(t)}$ a.a. $t \in T$, where δ_r designates the Dirac measure concentrated at point $r \in \mathbb{R}^k$. It is clear that $\mathscr{D}(t; \mu) = \frac{1}{1+\omega(|v(t)|)}$, and hence, ν_t satisfies (7.11).

By taking into account the definition of $\bar{g}$, we have:

$$x(t) = x_0 + \int_{t_0}^{t} \left\langle \bar{g}(x, v, \varsigma), d\wp \right\rangle =$$

$$= x_0 + \int_{t_0}^{t} \int_{v \in V} \left\langle \frac{g(x(\varsigma), v, \varsigma)}{1 + \omega(|v(\varsigma)|)}, d\delta_{v(\varsigma)} \right\rangle (1 + \omega(|v(\varsigma)|)) d\varsigma =$$

$$= x_0 + \int_{t_0}^{t} g(x(\varsigma), v(\varsigma), \varsigma) d\varsigma.$$

So, the trajectories generated by both controls are the same.

In the case of bounded problem (i.e., when set V is compact), may some other trajectories among the ones from the Gamkrelidze extension, [6], be encountered in the current extension (7.7)–(7.10)? The answer is negative. Indeed, in view of (7.11), it may easily be deduced that the measure μ is absolutely continuous. Moreover, its density equals

$$\frac{d\mu}{d\ell} = \int_{V} \Omega(v) d\nu_t \cdot \left(\int_{V} \frac{1}{1 + \omega(|v|)} d\nu_t \right)^{-1},$$

and hence bounded. (Note that $\bar{V} = V$ as V is bounded.) Then, by virtue of the given definitions, the set of feasible trajectories coincides with that of the conventional problem with generalized controls; see [6].

Therefore, new trajectories w.r.t. [6] may appear only when the set V is unbounded. Essentially, these new trajectories may exhibit discontinuities unlike those in the conventional problem with generalized controls, where they are absolutely continuous. On the basis of this, it may be concluded that the extension concept expressed by Problem (7.5) generalizes the approach from [6] concerning the case of discontinuous trajectories.

7.4 Generalized Existence Theorem

As was emphasized in the Introduction, the main objective of any extension undertaken in the optimal control theory is to ensure the existence of solution to the extended problem. The following existence theorem demonstrates consistency of Extensions (7.7)–(7.10) in this sense. In this section, for the sake of convenience, we consider that set S is a Cartesian product: $S = A \times B$, where A, B are some closed subsets of $\mathbb{R}^n$. This case is relevant w.r.t. to the majority of applications and, in particular, w.r.t. Examples (7.1)–(7.4). The general case is treated similarly.

Theorem 7.1 *Suppose that the control problem (7.5) allows the impulsive extension of order ω. Let $S = A \times B$, where at least one set A or B is compact, let $\mathscr{C} \neq \varnothing$, and assume that there exists a constant $\kappa > 0$ such that for any feasible control process $(x, \wp) \in \mathscr{C}$, where $\wp = \{\mu, \nu_t, \nu_s^\tau\}$, the following estimate is satisfied:*

$$\|\mu\| \leq \kappa. \tag{7.12}$$

Then, Problem (7.7) has a solution.

Proof The main idea of the proof is to reduce Problem (7.7) to a conventional convexified control problem by using the discontinuous time variable change π. Along with compactification (Θ, B_1), consider compactification (Θ_ω, B_1), where

$$\Theta_\omega(v) = \frac{\omega(|v|) \cdot v}{|v|(1 + \omega(|v|))}, \quad \text{when } v \in \mathbb{R}^k \setminus \{0\}, \quad \Theta_\omega(0) = 0,$$

and $\Theta_\omega(v)$ is the identity map over the infinity sphere. It is a straightforward task to verify that both compactifications are equivalent in the sense that they define the same topology; see Exercise 7.2. The second compactification is introduced in the proof for the sake of convenience.

Consider the following optimal control problem with generalized controls of the conventional type:

$$\begin{aligned}
&\text{Minimize } \varphi(x_0, x_1)\\
&\text{subject to: } \dot{x} = \langle f(x, u, \chi), \xi_s \rangle,\\
&\qquad \dot{\chi} = \alpha, \text{ a.a. } s \in [0, s_1],\\
&\qquad x_0 = x(0) \in A, \; x_1 = x(s_1) \in B,\\
&\qquad \chi(0) = t_0, \; \chi(s_1) = t_1,\\
&\qquad l(x, \chi) \leq 0,\\
&\qquad \alpha(s) \in [0, 1],\\
&\qquad \alpha(s) + \langle |u|, \xi_s \rangle = 1 \text{ a.a. } s,\\
&\qquad \operatorname{supp}(\xi_s) \subseteq U \; \forall s \in [0, s_1].
\end{aligned} \tag{7.13}$$

Here:

$$\begin{aligned}
f(x, u, \chi) &:= \bar{g}(x, \Theta_\omega^{-1}(u), \chi),\\
U &:= \Theta_\omega(\bar{V}).
\end{aligned}$$

Time endpoint s_1 in (7.13) is not fixed, unlike in Problem (7.7) considered on the fixed time interval $[t_0, t_1]$. The control functions in (7.13) are $\alpha(s)$ and ξ_s, where α is a usual measurable control function and ξ_s is a generalized control with support in U. Note that, set U is compact since it is the image of compact $\bar{V}$ by virtue of the continuous map Θ_ω. Therefore, Problem (7.13) is a conventional autonomous convexified control problem with the free time endpoint s_1; see [6].

Next, we show that the two problems, (7.7) and (7.13), are equivalent. That is, for every process $(x, \wp) \in \mathscr{C}$ there exists a feasible process $(\tilde{x}, \chi, \alpha, \xi_s, s_1)$ of Problem (7.13), such that the values of the minimizing functional are equal and vice versa.

Let $(x, \wp) \in \mathscr{C}$, where $\wp = \{\mu, \nu_t, \nu_s^\tau\}$. Consider the discontinuous time variable change $\pi(t)$. The inverse function of π is $\theta(s) : [0, s_1] \to T$, where $s_1 = t_1 - t_0 + \|\mu\|$.

Let us consider

$$\alpha(s) = \begin{cases} \mathscr{D}(\theta(s); \mu), & \text{when } s \in H(\mu), \\ 0, & \text{otherwise}, \end{cases}$$

$$\xi_s = \begin{cases} \vartheta_{\theta(s)}, & \text{when } s \in H(\mu), \\ \vartheta^\tau_{\zeta_\tau(s)}, & \text{otherwise}, \end{cases}$$

$$\tilde{x}(s) = \begin{cases} x(\theta(s)), & \text{when } s \in H(\mu), \\ x_\tau(\zeta_\tau(s)), & \text{otherwise}, \end{cases}$$

where $\vartheta_t, \vartheta_s^\tau$ are generalized controls on U, such that

$$\vartheta_t(E) = \nu_t(\Theta_\omega^{-1}(E)) \;\; \forall\, E \in \mathscr{B}(U),$$
$$\vartheta_s^\tau(E) = \nu_s^\tau(\Theta_\omega^{-1}(E)) \;\; \forall\, E \in \mathscr{B}(U),$$

and it is also defined:

$$H(\mu) := [0, s_1] \setminus \bigcup_{\tau \in \mathrm{Ds}(\mu)} \Gamma_\tau, \quad \zeta_\tau(s) := \frac{s - \pi(\tau^-)}{\ell(\Gamma_\tau)} : \Gamma_\tau \to [0, 1].$$

Note that, definitions of $\vartheta_t, \vartheta_{s,\tau}$ are correct in view of $\mathscr{B}(U) = \Theta_\omega(\mathscr{B}(\bar{V}))$.

Let us show that

$$\theta(s) = t_0 + \int_0^s \alpha(\varsigma) d\varsigma. \tag{7.14}$$

Indeed, by definition, we have

$$\ell([t_0, t]) = t - t_0 = \int_{t_0}^{t} \mathscr{D}(\sigma; \mu) d\pi(\sigma) = \int_0^{\pi(t)} \mathscr{D}(\theta(\varsigma); \mu) d\varsigma.$$

By taking into account (7.6), the definitions of α, θ, and also the fact that $\pi(\theta(s)) = s$ whenever $\pi(t)$ is continuous at point $t = \theta(s)$, and by considering in the above equality $t = \theta(s)$, we obtain (7.14). Then, $\chi = \theta$.

It is simple to verify that, due to (7.11) and the definition of Θ_ω, the constraint $\alpha(s) + \langle |u|, \xi_s \rangle = 1$ imposed in (7.13) is satisfied for a.a. s. It is also clear that the endpoint constraints and state constraints in (7.13) are satisfied.

By applying the variable change in (7.8), taking into account the trajectory concept, it follows that the arc $\tilde{x}(\cdot)$ satisfies the dynamical system in (7.13) together with the collection $(x_0, \alpha, \xi_s, s_1)$. Indeed,

$$\int_{t_0}^{t_1} \left\langle \bar{g}(x, v, t), d\wp \right\rangle = \int_0^{s_1} \left\langle f(\tilde{x}(s), u, \chi(s)), \xi_s \right\rangle ds.$$

Then, the above-constructed process $(\tilde{x}, \chi, \alpha, \xi_s, s_1)$ is feasible in (7.13), while the values of the cost function coincide.

Conversely, take a feasible process $(\tilde{x}, \chi, \alpha, \xi_s, s_1)$ of Problem (7.13). Regard the function $\chi(s)$ as an inverse function to a certain discontinuous time variable change $\tilde{\pi} : T \to [0, s_1]$. Function $\tilde{\pi}(t)$ is defined uniquely as such a function that $\tilde{\pi}(\chi(s)) = s$, a.a. s: $\alpha(s) > 0$, $\tilde{\pi}(t_0) = 0$, $\tilde{\pi}(t_1) = s_1$, and $\tilde{\pi}(t)$ is right-continuous on (t_0, t_1). Once $\tilde{\pi}$ is determined, we obtain the formula for μ:

$$\mu([t_0, t]) = \int_0^{\tilde{\pi}(t)} (1 - \alpha(s))ds.$$

Consider $v_t = \tilde{\xi}_{\tilde{\pi}(t)}$, and $v_s^\tau = \tilde{\xi}_{\tilde{\zeta}_\tau(s)}$, where

$$\tilde{\xi}_s(E) = \xi_s(\Theta_\omega(E)) \;\; \forall\, E \in \mathscr{B}(\bar{V}),$$

$$\tilde{\zeta}_\tau(s) = \ell(\tilde{\Gamma}_\tau)s + \tilde{\pi}(\tau^-) : [0, 1] \to \tilde{\Gamma}_\tau,$$

$$\tilde{\Gamma}_\tau = [\tilde{\pi}(\tau^-), \tilde{\pi}(\tau^+)], \quad \tau \in \mathrm{Ds}(\mu).$$

Now, the arc $x(t) = \tilde{x}(\tilde{\pi}(t)), t \in T$, and $x_\tau(s) = \tilde{x}(\tilde{\zeta}_\tau(s))$ is the solution to (7.8). It immediately follows from the variable change that the endpoint and state constraints are satisfied. So, the constructed process $(x, \wp)$, where $\wp = \{\mu, v_t, v_s^\tau\}$, is feasible in Problem (7.7), and the cost function takes the same value.

Thus, it has been demonstrated that the two problems, (7.7) and (7.13), are equivalent. Moreover, for the just established one-to-one correspondence it holds that

$$s_1 = t_1 - t_0 + \|\mu\|. \tag{7.15}$$

So, if a solution exists to one of the problems, then it also exists to the other. Let us ensure that the solution exists in the auxiliary problem (7.13). Note that this problem is not a standard problem with generalized controls due to the extra constraints imposed on α, ξ_s. However, the arguments proving its existence are standard and are, in essence, the same as in [6]. Let us briefly outline these arguments.

Since $\mathscr{C} \neq \varnothing$, the set of feasible processes in Problem (7.13) is not empty either. Consider a minimizing sequence $(x_i, \chi_i, \alpha_i, \xi_{s,i}, s_{1,i})$. The set of generalized controls is weakly sequentially compact, [6], so there exists a generalized control ξ such that, after passing to a subsequence, $\xi_i \xrightarrow{w} \xi$. Similarly, since the unit ball in L_2 is weakly sequentially compact, after passing to a subsequence, $\alpha_i \xrightarrow{w} \alpha$. The notation $\xrightarrow{w}$ stands for the weak convergence of controls.

From (7.12) and (7.15), it obviously follows that

$$0 < t_1 - t_0 \le s_{1,i} \le \text{const} .$$

Passing to a subsequence $s_{1,i} \to s_1 > 0$, $x_{0,i} \to x_0$ (or $x_{1,i} \to x_1$, depending on which set, A or B, is compact). The trajectory (x, χ) corresponding to (x_0, α, ξ_s) exists on the entire time interval $[0, s_1]$ due to the conditions imposed in Hypothesis (H1). Moreover, due to the weak convergence,

$$x_i(s) = x_{0,i} + \int_0^s \Big\langle f(x_i, u, \chi_i), \xi_{s,i} \Big\rangle ds \rightrightarrows x(s) = x_0 + \int_0^s \Big\langle f(x, u, \chi), \xi_s \Big\rangle ds,$$

and $x_i(s_{1,i}) \to x(s_1)$. By virtue of the weak convergence, we also have $\chi_i \rightrightarrows \chi$. Then, (x, χ) satisfies the endpoint and state constraints. Finally, by taking into account the definitions of weak convergence for generalized controls and for L_2-functions,[4] and in view of linearity, it is a straightforward task to verify that the constraint $\alpha(s) + \big\langle |u|, \xi_s \big\rangle = 1$ a.a. s is satisfied. Therefore, the collection $(x, \chi, \alpha, \xi_s, s_1)$ represents a solution to (7.13) as the sequence $(x_i, \chi_i, \alpha_i, \xi_{s,i}, s_{1,i})$ is minimizing. This implies the existence of solution to the original problem (7.7).

The proof is complete. □

Let us return to Examples (7.1)–(7.4) given in Sect. 7.1 and apply the just derived theorem. However, a few remarks need to be made first. In that which follows in this section, we consider that $A = \{x_0\}$. Then, $\varphi(x_0, x_1) = \varphi(x_1)$. This is assumed for the sake of convenience and, clearly, is not restrictive w.r.t. the above-mentioned examples. The general case is treated similarly.

Remark 7.1 Condition (7.12) can be replaced by

$$(x, \wp) \in \mathscr{C} \quad \Rightarrow \quad \|\mu\| \le r(\varphi(x_1)), \tag{7.16}$$

where $r(\cdot) : \mathbb{R} \to \mathbb{R}$ is some continuous increasing function. Indeed, in Problem (7.7), it is always possible to impose an additional and not restrictive endpoint constraint $\varphi(x_1) \le C$, where C is a sufficiently large number. Then, (7.12) is a corollary of (7.16).

Remark 7.2 In view of the fact that any Borel measure can be weakly-* approximated by absolutely continuous measures, and in view of the Gamkrelidze approximation lemma (see in [6]),[5] by using the discontinuous time variable change (as in the proof of Theorem 7.1) and the definition of generalized impulsive control $\wp$, it is

---

[4]Note that, these are two different types of convergence; see [6].

[5]This lemma, in particular, says that any generalized control can be weakly approximated by conventional controls, that is, in the sense of weak convergence of generalized controls.

not difficult to establish that for any $(x, \wp) \in \mathscr{C}$ there exists a sequence of control processes (x_i, v_i) of Problem (7.5), not necessarily feasible in (7.5), such that

$$\int_{t_0}^{t_1} \omega(|v_i(t)|)dt \to \|\mu\|,$$

$$x_i(t) \to x(t) \ \ \forall t \in (T \setminus \mathrm{Ds}(\mu)) \cup \{t_0\} \cup \{t_1\}, \ \text{ and}$$

$$\limsup_{i\to\infty} \max_{t\in T} l^j(x_i(t), t) \le 0 \ \ \forall\, j = 1, \dots, d.$$

By virtue of this property, Condition (7.12) will be satisfied, as soon as there are constants $\kappa, \delta > 0$ such that the estimate

$$\int_{t_0}^{t_1} \omega(|v(t)|)dt \le \kappa, \tag{7.17}$$

or, according to Remark 7.1, the estimate

$$\int_{t_0}^{t_1} \omega(|v(t)|)dt \le r(\varphi(x_1)), \tag{7.18}$$

is valid for any control function $v : T \to V$, for which

$$\mathrm{dist}(x(t_1), B) \le \delta, \ \ l^j(x(t), t) \le \delta, \ \ j = 1, \dots, d, \ \ t \in T,$$

where $\dot{x}(t) = g(x(t), v(t), t)$, and $x(t_0) = x_0$.

Therefore, for the verification of (7.12) or (7.16) it is not obligatory to pass to the class of generalized impulsive controls. At the same time, the above sufficient condition for validity of (7.12) or (7.16) is often simple to verify. In particular, as we will now see, it holds true for Examples (7.1)–(7.4).

Now, by virtue of Remark 7.2, let us ensure that the extension expressed by (7.7)–(7.10) is well posed for Problems (7.1)–(7.4) in the sense that the solution in the extended problem exists. Regarding the Dido problem (7.2) and Catenary (7.3), it is rather clear, as by setting $\omega(\rho) = \rho$, $v = \dot{x}$, $g^1(x, v) := v$, $g^2(x, v) = \sqrt{1 + v^2}$, for any $\delta > 0$, we have

$$\int_0^1 |v|dt \le \int_0^1 \sqrt{1 + |v|^2}dt = \int_0^1 g^2(x, v)dt \le L + \delta.$$

Then, (7.17) with $\kappa = L + \delta$ guarantees the existence of solution.

For Problem (7.4), Condition (7.17) is useful once more. We should consider $\omega(\rho) = \rho^2, g^1(x, v) := v, g^2(x, v) = v^2, \kappa = 1 + \delta$, in order to justify the existence of solution.

Regarding Example (7.1), the existence of solution is also obvious if the additional constraint $x(t) \geq \varepsilon$, where $\varepsilon > 0$, is imposed. Indeed, this is due to Condition (7.18) and the following estimate (here, $\omega(\rho) = \rho$)

$$\varepsilon \int_0^1 |v| dt \leq \int_0^1 \varepsilon \sqrt{1 + |v|^2} dt \leq \int_0^1 x \sqrt{1 + |\dot{x}|^2} dt.$$

The case $x(t) \geq 0$ requires additional considerations. It does not follow directly from Theorem 7.1, but can be proved by the ε-approximations as $\varepsilon \to 0$.

7.5 Maximum Principle

In this section, the necessary conditions of optimality in the form of Pontryagin maximum principle for problem (7.7) are presented.

Let us introduce necessary notations. Let

$$\bar{H} : \mathbb{R}^n \times \bar{\mathbb{R}}^k \times \mathbb{R}^n \times \mathbb{R}^l \times \mathbb{R}^1 \times \mathbb{R}^1 \to \mathbb{R}^1$$

be the extended Hamilton–Pontryagin function:

$$\bar{H}(x, v, \psi, \gamma, t) := \langle \psi, \bar{g}(x, v, t) \rangle - \langle \gamma, \Gamma(x, v, t) \rangle,$$

where

$$\Gamma(x, v, t) := \frac{\partial l}{\partial x}(x, t)\bar{g}(x, v, t) + \frac{\partial l}{\partial t}(x, t)$$

is the derivative w.r.t. the control system. (This function was first proposed in [3], see also [7], for use in the study of state constraints.) The set of all generalized controls $\wp = \{\mu, \nu_t, \nu_s^\tau\}$ defined on $\bar{V} \times T$ is denoted by $\mathscr{P}$.

Theorem 7.2 *Let $(\hat{x}, \hat{\wp})$, where $\hat{\wp} = (\hat{\mu}, \hat{\nu}_t, \hat{\nu}_s^\tau)$, be an optimal process in Problem (7.7). Assume that Hypotheses (H1), (H2) are in force.*

Then, there exist a number $\lambda \geq 0$, a vector-valued function of bounded variation ψ, together with its attached family ψ_τ, and a decreasing function γ, together with its attached family γ_τ, where $\tau \in \mathrm{Ds}(\hat{\mu})$, such that:

(C1) *Pair $(\hat{x}, \psi)$ satisfies the generalized Hamiltonian system:*

$$\begin{cases} d\hat{x} = \left\langle \frac{\partial \bar{H}}{\partial \psi}(\hat{x}, v, \psi, \gamma, t), d\hat{\wp} \right\rangle \\ d\psi = -\left\langle \frac{\partial \bar{H}}{\partial x}(\hat{x}, v, \psi, \gamma, t), d\hat{\wp} \right\rangle, \end{cases}$$

on time interval $[t_0, t_1]$, *which implies that for attached families* $\hat{x}_\tau, \psi_\tau, \tau \in \mathrm{Ds}(\hat{\mu})$, *given on* $[0, 1]$, *it holds that*

$$\begin{cases} \dot{\hat{x}}_\tau = \Delta_\tau \cdot \left\langle \frac{\partial \bar{H}}{\partial \psi}(\hat{x}_\tau, v, \psi_\tau, \gamma_\tau, \tau), \hat{v}_s^\tau \right\rangle \\ \dot{\psi}_\tau = -\Delta_\tau \cdot \left\langle \frac{\partial \bar{H}}{\partial x}(\hat{x}_\tau, v, \psi_\tau, \gamma_\tau, \tau), \hat{v}_s^\tau \right\rangle, \end{cases}$$

and also, $\hat{x}_\tau(0) = \hat{x}(\tau^-)$, $\psi_\tau(0) = \psi(\tau^-)$;

(C2) *Transversality conditions*

$$\left(\psi(t_0) - \frac{\partial l}{\partial x}(\hat{x}_0, t_0)\gamma(t_0), -\psi(t_1)\right) \in \lambda\varphi'(\hat{p}) + N_S(\hat{p}),$$

where $\hat{x}_0 = \hat{x}(t_0)$;

(C3) *Maximum condition*

$$\max_{\wp \in \mathscr{P}: \|\mu\| = \|\hat{\mu}\|} \int_{t_0}^{t_1} \left\langle \bar{H}(\hat{x}, \psi, v, \gamma, t), d\wp \right\rangle = \int_{t_0}^{t_1} \left\langle \bar{H}(\hat{x}, \psi, v, \gamma, t), d\hat{\wp} \right\rangle;$$

(C4) *Nontriviality condition*

$$\lambda + |\psi(t_0)| + |\gamma(t_0)| > 0.$$

Moreover, $\gamma(t), \gamma_\tau(s), \tau \in \mathrm{Ds}(\hat{\mu})$, *satisfy the following conditions:*

(a) *Functions* γ^j, γ_τ^j, $j = 1, \ldots, d$, *are constant on any time interval in* T, $[0, 1]$, *where optimal paths* $\hat{x}(t), \hat{x}_\tau(s)$ *lie in the interior of the* j^{th} *state constraint set, respectively.*

(b) *Functions* γ^j *are left continuous on the interval* (t_0, t_1), *and* γ_τ^j *are left continuous on the interval* $(0, 1)$.

(c) *Functions* γ^j *are decreasing,* $\gamma^j(t_1) = 0$, γ_τ^j *are decreasing, and* $\gamma_\tau(0) = \gamma(\tau^-)$, $\gamma_\tau(1) = \gamma(\tau^+)$.

Above, for brevity: in (C1), (C3) the dependence of the extremal on t, s is omitted. The solutions to (C1) and the meaning of the integral in (C3) are understood as before, via the attached families; see Sect. 7.3.

The proof of Theorem 7.2 is rather straightforward: It can be organized by implementing the discontinuous time variable change as has been carried out in Chap. 4 and by using the results from [1]. Note that the maximum principle from Theorem 7.2 degenerates: That is, it can be satisfied by a trivial collection of Lagrange multipliers as soon as one of the endpoints $\hat{x}_0, \hat{x}_1$ lies within the boundary of the state constraint set. To achieve nondegenerate conditions, one may use the technique proposed in Chap. 5; see also [2].

7.6 Examples of Extension

In this section, the extensions for Examples (7.1)–(7.4) in the framework of the approach of Sect. 7.3 are presented. Begin with the extended solutions to the minimal surface problem (7.1).

Euler problem. The classical continuous solution is well known—the optimal arc is given by the hyperbolic cosine: $x(t) = \mathrm{ch}(t - 1/2)$. (Here, we set $r_1 = r_2 = (e^{-1/2} + e^{1/2})/2$.) The discontinuous solution, when the numbers r_1, r_2 are sufficiently small, is also clear. Indeed, as has been mentioned in Sect. 7.1, it is given by the discontinuous arc $x(0) = r_1$, $x(1) = r_2$ $x(t) = 0$, $t \in (0, 1)$. However, according to the approach proposed here, the discontinuous solution *is not only* the discontinuous trajectory $x(t)$ itself, but also the attached family of continuous arcs $x_\tau(s)$ defined over $[0, 1]$, specifying the behavior of the trajectory along the jumps at points τ.

Let us describe the precise extension by following (7.7). The Euler problem involves minimizing of the integral with $g_0(x, v) = x\sqrt{1 + v^2}$, where $v = \dot{x}$,—the formulation which is covered by (7.5); see Sect. 7.1. By setting $\omega(\rho) = \rho$, in view of the definition, we have

$$\bar{g}(x, v) = \frac{v}{1 + |v|}, \quad \bar{g}_0(x, v) = x\frac{\sqrt{1 + v^2}}{1 + |v|},$$

and $\bar{g}(x, v) = \pm 1$, $\bar{g}_0(x, v) = x$, when $v \in S_\infty$ (i.e., when $v = \pm\infty$). Then, the optimal generalized impulsive control $\wp$ is:

$$\mu = r_1\delta_0 + r_2\delta_1, \quad \nu_t = 0, \quad \nu_s^0 = \delta_{-\infty}, \quad \nu_s^1 = \delta_{+\infty}.$$

The optimal extended trajectory is:

$$x(t) = \begin{cases} r_1, & t = 0, \\ 0, & t \in (0, 1), \\ r_2, & t = 1, \end{cases} \qquad x_0(s) = r_1 - r_1 s, \quad x_1(s) = r_2 s.$$

Note that the attached functions $x_0(s)$, $x_1(s)$ demonstrate the linear character of the rupture at the endpoints.

The form of this solution, on the one hand, follows from simple geometrical considerations. On the other hand, it is a straightforward exercise to ensure that this solution satisfies the maximum principle stated in Theorem 7.2. Moreover, there are only a finite number of extremals (in fact, only two) which satisfy the necessary optimality conditions and which are simple to calculate; see Exercise 7.3. (On the basis of this, the conclusion is drawn that the presented extended trajectory yields the solution indeed.)

Regarding this extension of (7.1), it is also important to mention the following. Firstly, note that the value of integral in (7.1) is not exactly the area of the minimal surface, the integral needing to be multiplied by 2π. Let us compute the value of this integral on the discontinuous solution indicated above. By definition, we have:

$$2\pi \int_0^1 \bar{g}_0(x,v)d\wp = 2\pi \int_0^1 r_1 x_0(s)ds + 2\pi \int_0^1 r_2 x_1(s)ds =$$

$$= 2\pi \int_0^1 r_1^2(1-s)ds + 2\pi \int_0^1 r_2^2 s ds = \pi r_1^2 + \pi r_2^2.$$

Thus, the answer is the area of two disks with radiuses r_1, r_2. This is fully consistent with the physical interpretation of the problem. Indeed, the minimal surface in the discontinuous case consists of the sides of the disks facing each other plus the connecting segment along the axis [0, 1]. The area of the segment is zero, while the area of the disks is exactly the above value $\pi r_1^2 + \pi r_2^2$.

Dido problem. In the Dido problem (7.2), when $L \le \pi$, the solution is classical—it is the arc of the circle, connecting two points $t = -1$ and $t = 1$. When $L > \pi$, the solution is discontinuous. Let us describe it.

By definition, for $\omega(\rho) = \rho$, we have:

$$\bar{g}(x,v) = \frac{v}{1+|v|}, \quad \bar{g}_0(x,v) = \frac{x}{1+|v|},$$

and $\bar{g}(x,v) = \pm 1$, $\bar{g}_0(x,v) = 0$, when $v \in S_\infty$ (i.e., when $v = \pm\infty$). Then, the optimal generalized impulsive control $\wp$ is:

$$\mu = r\delta_{-1} + r\delta_1 + \mu_a, \quad \nu_t = \delta_{a(t)}, \quad \nu_s^{-1} = \delta_{+\infty}, \quad \nu_s^1 = \delta_{-\infty},$$

where $r = \frac{L-\pi}{2}$, μ_a is an absolutely continuous measure such that $\frac{d\mu_a}{dt} = |a(t)|$, and

$$a(t) = \frac{-t}{\sqrt{1-t^2}}.$$

The optimal trajectory $x(t)$ is:

$$x(t) = \begin{cases} 0, & t = -1, \\ r + \sqrt{1-t^2}, & t \in (-1,1), \\ 0, & t = 1, \end{cases}$$

and $x_0(s) = rs$, $x_1(s) = r - rs$.

Once again, by applying the maximum principle from Theorem 7.2, one can ensure that the constructed arcs are the solutions indeed.

Catenary. The construction repeats, in essence, that of the Euler problem. This extension is given as an exercise; see Exercise 7.4.

Problem (7.4). Regarding Problem (7.4), the extended solution is trivial. The number of solutions, in fact, is infinite. For example,

$$\mu = \delta_\tau, \ \nu_t = 0, \ \nu_s^\tau = \delta_\infty, \ \text{or} \ \nu_s^\tau = \delta_{-\infty},$$

where $\tau \in [0, 1]$ is chosen arbitrarily. At the same time, $\omega(\rho) = \rho^2$.

As can be seen, Examples (7.1)–(7.4) considered earlier possess extensions in the framework of the above proposed approach.

7.7 Exercises

Exercise 7.1 Ensure that the topology base in $\bar{\mathbb{R}}^m$ is constituted by the open sets in $\mathbb{R}^m$, plus the following sets

$$\Big(\big(\mathbb{R}^m \setminus \operatorname{cl} O_{1/\varepsilon}(0)\big) \cap \operatorname{cone} O_\varepsilon(3 \cdot \Theta^{-1}(l/2))\Big) \cup \Big(O_\varepsilon(l) \cap S_\infty\Big), \quad \varepsilon > 0, \ l \in S_\infty,$$

where $O_R(x)$ is the open ball in $\mathbb{R}^m$ of radius R centered at x.

Exercise 7.2 Ensure the equivalence of the compactifications given by Θ and Θ_ω (see the proof of Theorem 7.1).

Exercise 7.3 By applying the maximum principle from Theorem 7.2 to the extended Euler problem, find all the extremals. Ensure that the discontinuous solutions do exist.

Exercise 7.4 Construct the extension for catenary problem (7.3).

Exercise 7.5 Condition (7.11) is important. Without it, the suggested formulation in (7.7) is not a well-posed extension of (7.5). Ensure the validity of this assertion with the simplest examples. That is, in the absence of (7.11), construct an example of a feasible extended trajectory such that $x(t_1) \notin \operatorname{cl} \mathscr{A}(t_1)$, where $\mathscr{A}(t)$ is the reachable set of the original dynamical system at time t. (Then, such extended trajectory cannot be approximated by the trajectories of the original problem.)

Exercise 7.6 Return to the model example of Chap. 6, Sect. 6.2. Imagine that in this example the resulting propulsion force depends nonlinearly on the fuel consumption rate v that which is a realistic assumption. Therefore, suppose that the propulsion force is proportional not to v, but to some function ϕ of v which is nonlinear and continuous and may characterize the efficiency of the fuel consumption. Regarding function ϕ, assume that

$$\lim_{v \to \infty} \frac{\phi(v)}{v} = \kappa,$$

where κ is some positive number. Consider, for example, that $\phi(v) = \kappa\sqrt{v^2 - 1}$, $v \geq 1$ and $\phi(v) = 0$ when $v \in [0, 1]$.

Construct the extension of the minimal fuel consumption problem in this nonlinear case. Ensure the existence of solution by virtue of Theorem 7.1. Simplify the equations of motion by considering the disk motion without rotation, that is, set $\alpha = 0$. (Then, clearly, the control parameter θ is redundant as $\theta_i = \theta_i(m)$, $i = 1, 2, 3, 4$.)

Exercise 7.7* Note that in the extensions given for the classic examples (7.1)–(7.3) the generalized impulsive control always appears as the family of Dirac measures and, thereby, could be rewritten in terms of conventional measurable functions $v(t)$, $v_\tau(s)$ and Borel measure μ. This fact, as a matter of principle, raises the question of a more accurate extension for (7.5). "More accurate," in this context, means that the extension is reduced as it contains less generalized impulsive controls and less extended trajectories. This can be achieved by dispensing with the convexification procedure. The price of such an approach is a weaker existence theorem which may not always be applicable and which would require more thorough hypotheses. Investigate this question and construct such an extension, bearing in mind the following condition to be imposed on the components of the dynamics (verify that for (7.1)–(7.3) this condition is valid):

$$g^j(x, \lambda v, t) \le \lambda g^j(x, v, t), \quad \forall \lambda \ge 1.$$

References

1. Arutyunov, A., Karamzin, D., Pereira, F.: The maximum principle for optimal control problems with state constraints by R.V. Gamkrelidze: revisited. J. Optim. Theory Appl. **149**(3), 474–493 (2011)
2. Arutyunov, A.V.: Optimality conditions. Abnormal and degenerate problems. In: Mathematics and its Applications. Kluwer Academic Publishers, Dordrecht (2000)
3. Gamkrelidze, R.: Optimal control processes for bounded phase coordinates. Izv. Akad. Nauk SSSR. Ser. Mat. **24**, 315–356 (1960)
4. Gamkrelidze, R.: On sliding optimal states. Soviet Math. Dokl. **3**, 390–395 (1962)
5. Gamkrelidze, R.: On some extremal problems in the theory of differential equations with applications to the theory of optimal control. J Soc. Ind. Appl. Math. Ser. A Control **3**(1), 106–128 (1965)
6. Gamkrelidze, R.: Principles of Optimal Control Theory. Plenum Press, New York (1978)
7. Pontryagin, L., Boltyanskii, V., Gamkrelidze, R., Mishchenko, E.: Mathematical theory of optimal processes. Translated from the Russian ed. by L.W. Neustadt. Interscience Publishers, Wiley, 1st edn (1962)
8. Warga, J.: Optimal Control of Differential and Functional Equations. Academic Press, New York, London (1972)

Correction to: Impulsive Control Problems Without the Frobenius Condition

Correction to: Chapter 4 in: A. Arutyunov et al., *Optimal Impulsive Control*, Lecture Notes in Control and Information Sciences 477, https://doi.org/10.1007/978-3-030-02260-0_4

In the original version of the book, the following belated correction has been incorporated in Chapter 4. In Theorem 4.1, the condition (b) has been changed from '*The set U and cone K are convex.*' to '*The set f(x, U, t) and cone K are convex.*' The correction chapter and the book have been updated with the change.

The updated version of this chapter can be found at https://doi.org/10.1007/978-3-030-02260-0_4

© Springer Nature Switzerland AG 2019

A. Arutyunov et al., *Optimal Impulsive Control*, Lecture Notes in Control and Information Sciences 477, https://doi.org/10.1007/978-3-030-02260-0_8

Index

© Springer Nature Switzerland AG 2019

A. Arutyunov et al., *Optimal Impulsive Control*, Lecture Notes in Control and Information Sciences 477, https://doi.org/10.1007/978-3-030-02260-0

The manufacturer's authorised representative in the EU is Springer Nature Customer Service Centre GmbH, Europaplatz 3, 69115 Heidelberg, Germany. If you have any concerns regarding our products, please contact ProductSafety@springernature.com

Printed and bound by CPI Group (UK) Ltd, Croydon, CR0 4YY
05/10/2026
02228145-0001